# ENCYCLOPEDIA OF NANOTECHNOLOGY

# ENCYCLOPEDIA OF
# NANOTECHNOLOGY

*By*
Rakesh Rathi

2009

SBS Publishers & Distributors Pvt. Ltd.
New Delhi

ISBN 13 : 9789380090054

First Published in India in 2009



Published by:

SBS PUBLISHERS & DISTRIBUTORS PVT. LTD.
2/9, Ground Floor, Ansari Road, Darya Ganj,
New Delhi - 110002,
INDIA
Tel: 0091.11.23289119 / 41563911 / 32945311
Email: mail@sbspublishers.com
www.sbspublishers.com

Printed in India by Chaman Enterprises, New Delhi.

# Preface

This encyclopaedia presents a comprehensive list of terms used in the "Encyclopaedia of Nanotechnology" and various topics related with it. The content is a natural reflection of the spread of this subject. This vital reference offers a wealth of essential information in a portable, convenient, quick-find format. Whether reader is a professional or a student, there is no better way to stay knowledgeable about current terminology of the various branches of 'Nanotechnology' and gain an understanding of their key ideas and concepts.

This encyclopaedia has been written in a clear and simple language, easily understood by the general reader. In addition, it provides an in-depth understanding of the subject, thus providing enough information for educators, scientists, advanced students and corporates. The style is easy to read and suitable for non-native English speakers and translators with no science or engineering background.

The work is designed to give readers direct insight into the main error sources occurring in their profession, especially those resulting from a poor understanding of the subject matter and the usage of particular terms to designate different concepts in different branches of 'Nanotechnology'.

This work includes the words, phrases, acronyms and other abbreviations that used by those who work and write in these fields. The words may be either those used uniquely in a field or more common words that have a special meaning in the context of the 'Nanotechnology'.

Carefully reviewed for clarity, completeness and accuracy, this encyclopaedia offers a standard of excellence unmatched by any similar publication.

**Rakesh Rathi**

# A

## Accelrys

Accelrys (NASDAQ: ACCL) - Accelrys is the leader in Scientific Business Intelligence and Innovation that accelerates the discovery and development of novel therapeutics and materials. Accelrys develops and commercializes Scientific Business Intelligence software and solutions for the integration, mining, analysis, modeling, simulation, management and interactive reporting of scientific data. It provides solutions and products to Life Sciences and Material Sciences Industries, Government, and Academia worldwide.

Accelrys manages the Nanotechnology Consortium preoduces software tools for rational nanodesign. Accelrys has also created a SIG to develop the first ever biological registration system. The system is designed to capture IP for the biological entity and allow the users to uniquely identify each testable entity. Accelrys runs several SIGs focusing on automation and parameter generation.

## Acceptor (semiconductors)

In semiconductors physics the term acceptor is used to generically indicate a dopant atom that added to a semiconductor can form p-type regions.For example, when Silicon (Si), having four valence electrons, needs to be doped as an p-type semiconductor, elements from group III like Boron (B) or Aluminium (Al) have three valence electrons. When substituting a B atom in the crystalline lattice the three valence electrons form covalent bonds with three of the Si neighbours but the bond with the fourth neighbour remains unsatisfied. The unsatisfied bond attracts electrons from the neighbouring bonds. At room temperature always an electron from a neighbouring bond will jump to repair the unsatisfied bond leaving a hole which can move around the crystal and carry a current. The initially electroneutral acceptor becomes negatively charged (ionised).

## Accuracy

In the fields of science, engineering, industry and statistics, accuracy is the degree of closeness of a measured or calculated quantity to its actual (true) value. Accuracy is closely related to precision, also called reproducibility or repeatability, the degree to which further measurements or calculations show the same or similar results. The results of calculations or a measurement can be accurate but not precise; precise but not accurate; neither; or both. A measurement system or computational method is called *valid* if it is both *accurate* and *precise.* The related terms are *bias* (non-random or directed effects caused by a factor or factors unrelated by the independent variable) and *error* (random variability), respectively .

Accuracy is the degree of veracity while precision is the degree of reproducibility. The analogy used here to explain the difference between accuracy and precision is the target comparison. In this analogy, repeated measurements are compared to arrows that are fired at a target. Accuracy describes the closeness of arrows to the bullseye at the target center. Arrows that strike closer to the bullseye are considered more accurate. The closer a system's measurements to the accepted value, the more accurate the system is considered to be. To continue the analogy, if a large number of arrows are fired, precision would be the size of the arrow cluster. (When only one arrow is fired, precision is the size of the cluster one *would* expect if this were repeated many times under the same conditions.) When all arrows are grouped tightly together, the cluster is considered precise since they all struck close to the same spot, if not necessarily near the bullseye. The measurements are precise, though not necessarily accurate.

However, it is *not* possible to reliably achieve accuracy in individual measurements without precision — if the arrows are not grouped close to one another, they cannot all be close to the bullseye. (Their *average* position might be an accurate estimation of the bullseye, but the individual arrows are inaccurate.) See also Circular error probable for application of precision to the science of ballistics.

Ideally a measurement device is both accurate and precise, with measurements all close to and tightly clustered around the known value. The accuracy and precision of a measurement process is usually established by repeatedly measuring some traceable reference standard. Such standards are defined in the International System of Units and maintained by national standards organizations such as the National Institute of Standards and Technology.

In many cases precision can be characterised in terms of the standard deviation of the measurements, sometimes incorrectly called the measurement process's standard error. The smaller the standard deviation, the higher the precision. In some literature, precision is defined as the reciprocal of variance, while many others still confuse precision with the confidence interval. The interval defined by the standard deviation is the 68.3% ("one sigma") confidence interval of the measurements. If enough measurements have been made to accurately estimate the standard deviation of the process, and if the measurement process produces normally distributed errors, then it is likely that 68.3% of the time, the true value of the measured property will lie within one standard deviation, 95.4% of the time it will lie within two standard deviations, and 99.7% of the time it will lie within three standard deviations of the measured value. This also applies when measurements are repeated and averaged. In that case, the term standard error is properly applied: the precision of the average is equal to the known standard deviation of the process divided by the square root of the number of measurements averaged. Further, the central limit theorem shows that the probability distribution of the averaged measurements will be closer to a normal distribution than that of individual measurements.

## Acetyl

In organic chemistry, acetyl (IUPAC name ethanoyl) is a functional group, the acyl of acetic acid, with chemical formula $-COCH_3$. It is sometimes abbreviated as Ac (not to be confused with the element actinium). The acetyl radical contains a methyl group single-bonded to a carbonyl. The carbon of the carbonyl has a lone electron available, with which it forms a chemical bond to the remainder *R* of the molecule.

The acetyl radical is a component of many organic compounds, including the neurotransmitter acetylcholine and the analgesics acetaminophen and acetylsalicylic acid (better known as aspirin). Acetyl groups are frequently added to histones and other proteins modifying their properties. The introduction of an acetyl group into a molecule is called acetylation (or ethanoylation). In biological organisms, acetyl groups are commonly transferred bound to Coenzyme A (CoA), in the form of acetyl-CoA. Acetyl-CoA is an important intermediate both in the biological synthesis and in the breakdown of many organic molecules.

## Acetylcholine Receptor

Ion channel that opens in response to acetylcholine binding, thereby converting a chemical signal into an electrical one. Best understood example of a ligand-gated channel. Sometimes called the *nicotinic* acetylcholine receptor to distinguish it from a *muscarinic* receptor, which is a G-protein-linked cell-surface receptor.

## Acetylcholine

Neurotransmitter that functions at cholinergic chemical synapses, found both in the brain and in the peripheral nervous system. It is the neurotransmitter at vertebrate neuromuscular junctions.

## Acrosomal Process

Long, thin, actin-containing spike produced from the head of certain sperm when they make contact with the egg. Seen in sea urchins and other marine invertebrates whose eggs are surrounded by a thick gelatinous coat.

## Acrosome

In spermatozoa (also known as sperm/sperm cell) of many animals, the Acrosome is an organelle that develops over the anterior half of the sperm's head. It is a cap-like structure derived from the Golgi apparatus. Acrosome formation is completed during testicular maturation. In Eutherian mammals the acrosome contains digestive enzymes (including hyaluronidase and acrosin). These enzymes breakdown the outer membrane of the ovum called the zona pellucida, allowing the haploid nuclei in the sperm to join with the haploid nucleus found in the ova.

## Actin Filament

Helical protein filament formed by the polymerization of globular actin molecules. A major constituent of the cytoskeleton of all eucaryotic cells and part of the con- tractile apparatus of skeletal muscle.

## Actin

Actin is a globular, roughly 42-kDa protein found in all eukaryotic cells (except for nematode sperm) where it may be present at concentrations of over 100¼ M. It is also one of the most highly-conserved proteins, differing by no more than 20% in species as diverse as algae and humans. It is the monomeric subunit of microfilaments, one of the three major components of the cytoskeleton, and of thin filaments, which are part of the contractile apparatus in muscle cells. Thus, actin participates in many important cellular functions, including muscle contraction, cell motility, cell division and cytokinesis, vesicle and organelle movement, cell signaling, and the establishment and maintenance of cell junctions and cell shape.

Principal interactions of structural proteins at cadherin-based adherens junction. Actin filaments are linked to á-actinin and to membrane through vinculin. The head domain of vinculin associates to E-cadherin via á-, â-, and ã-catenins. The tail domain of vinculin binds to membrane lipids and to actin filaments.

The protein actin is one of

the most highly conserved throughout evolution because it interacts with a large number of other proteins, with 80.2% sequence conservation at the gene level between *Homo sapiens* and *Saccharomyces cerevisiae* (a species of yeast), and 95% conservation of the primary structure of the protein product.

Although most yeasts have only a single actin gene, higher eukaryotes, in general, express several isoforms of actin encoded by a family of related genes. Mammals have at least six actin isoforms coded by separate genes, which are divided into three classes (alpha, beta and gamma) according to their isoelectric point. In general, alpha actins are found in muscle (á-skeletal, á-aortic smooth, á-cardiac, and ã2-enteric smooth), whereas beta and gamma isoforms are prominent in non-muscle cells (â- and ã1-cytoplasmic). Although the amino acid sequences and *in vitro* properties of the isoforms are highly similar, these isoforms cannot completely substitute for one another *in vivo*.

The typical actin gene has an approximately 100-nucleotide 5' UTR, a 1200-nucleotide translated region, and a 200-nucleotide 3' UTR. The majority of actin genes are interrupted by introns, with up to 6 introns in any of 19 well-characterised locations. The high conservation of the family makes actin the favoured model for studies comparing the introns-early and introns-late models of intron evolution.

All non-spherical prokaryotes appear to possess genes such as MreB, which encode homologues of actin; these genes are required for the cell's shape to be maintained. The plasmid-derived gene ParM encodes an actin-like protein whose polymerised form is dynamically unstable, and appears to partition the plasmid DNA into the daughter cells during cell division by a mechanism analogous to that employed by microtubules in eukaryotic mitosis. Actin is found in both smooth and rough endoplasmic reticulums.

## Actin-binding Protein

Actin-binding proteins (also

known as ABP) are proteins that bind to actin. This may mean ability to bind actin monomers, or polymers, or both. Many actin-binding proteins, including ±-actinin, ²-spectrin, dystrophin, utrophin and fimbrin, do this through the actin-binding calponin homology domain. This is a list of actin-binding proteins in alphabetical order.

## Activation Energy

Activation energy in chemistry and biology is the threshold energy, or the energy that must be overcome in order for a chemical reaction to occur. Activation energy may otherwise be denoted as the minimum energy necessary for a specific chemical reaction to occur. The activation energy of a reaction is usually denoted by $E_a$. Known as the collisional model, there are three necessary requirements in order for a reaction to take place: Molecules must collide to react. If two molecules simply collide, however, they will not always react; therefore, the occurrence of a collision is not enough. The second requirement is that there must be enough energy (energy of activation) for the two molecules to react. This is the idea of a transition state; if two slow molecules collide, they might bounce off one another because they do not contain enough energy to reach the energy of activation and overcome the transition state (the highest energy point). Lastly, the molecules must be oriented with respect to each other correctly. For the reaction to occur between two colliding molecules, they must collide in the correct orientation, and possess a certain, minimum, amount of energy.

As the molecules approach each other, their electron clouds repel each other. Overcoming this repulsion requires energy (activation energy), which is typically provided by the heat of the system; i.e., the translational, vibrational, and rotational energy of each molecule, although sometimes by light (photochemistry) or electrical fields (electrochemistry). If there is enough energy available, the repulsion is overcome and the molecules get close enough for attractions between the molecules to cause a rearrangement of bonds. At low temperatures for a particular reaction, most (but not all) molecules will not have enough energy to react. However there will nearly always be a certain number with enough energy at any temperature because temperature is a measure of the *average* energy of the system — individual molecules can have more or less energy than the average. Increasing the temperature increases the proportion of molecules with more energy than the activation energy, and consequently the rate of reaction increases. Typically the activation energy is given as the energy in kilojoules needed for one mole of reactants to react.

### Active Component

Non-mechanical circuit component that has gained or switched current flow, such as a diode, transistor, etc.

### Active Device

Device requiring a source of energy for its operation and has an output that is a function of present and past input signals. Examples include controlled power supplies, transistors, LEDs, amplifiers, and transmitters.

### Active Layer

A layer of a microelectronic device in which both electrons and holes are active. In MOSFETs, bulk, silicon is the active layer. In thin-film transistors, a thin silicon film is the active layer.

### Active Site

Region of an enzyme surface to which a substrate molecule binds in order to undergo a catalyzed reaction.

### Active Transport

Movement of a molecule across a membrane or other barrier driven by energy other than that stored in the concentration gradient or electrochemical gradient of the transported molecule.

### Acyl Group

Functional group derived from a carboxylic acid. (R represents an alkyl group, such as methyl.)

### Adsorption

Formation of molecules of a gas similar to a thin film that binds on the surface of a solid. Unlike absorption, the binding to the surface in adsorption is usually weak and reversible. Adsorption is a method for separating mixtures and purifying fluids.

**Affinity Constant**

The reciprocal of the dissociation constant; a measure of the binding energy of a ligand in a receptor.

**Affinity**

A thermodynamic measurement of the strength of binding between molecules, say between an antibody and antigen. Each antibody/antigen pair has an association constant, Ka, expressed in L/mol.

**Akebono**

A Japanese scientific studies satellite that became operational in 1989. It contained only one non-Japanese instrument: Canada's Suprathermal Ion Mass Spectrometer, sent to investigate the Earth's outer atmosphere.

**Algorithm**

A set of well-defined mathematical rules or operations for solving a problem in a finite number of steps.

**Aliasing**

A process whereby two or more frequencies, integral multiples of each other, cannot be distinguished from each other when sampled in an analog-to-digital converter.

**Amide**

A molecule containing an amine bonded to a carboxyl group; the resulting bond has substantial double-bond character. Also termed a peptide; amide bonds link amino acids in proteins.

**Amine**

A molecule containing N with a single bond to C and two other single bonds to H or C (but not an amide); the amine group or moiety.

**Amino Acid**

A molecule containing both an amine and a carboxylic acid group; in the 20 genetically encoded amino acids in biology, both groups are bound to the same C. Amino acids joined by amide bonds form peptides and proteins; these do not contain amino acids as such, and are instead said to contain *amino acid residues*. Organic molecules that are the building blocks of proteins. There are some two hundred known amino acids, of which twenty are used extensively in living organisms.

## Amperometric Sensor

Amperometric sensors involve a heterogeneous electron transfer as a result of an oxidation/reduction of an electroactive species at a sensing electrode surface. A current is measured at a certain imposed voltage of the sensing electrode with respect to the reference electrode. Analytical information is obtained from the current-concentration relationship at that given applied potential.

## Analogue Based Drug Discovery

Considerable success has been achieved in the development of new drugs based on the leads provided by established drugs. Such new drugs are often considered deprecatingly as "me-too" products. However the book will demonstrate that it is part of the very important process where by drug action is optimized for a given therapeutic effect.

## Anechoic Chamber

Also called an "echo free" chamber. A room at the David Florida Laboratory where satellites are tested to make sure they are sending and receiving clear signals. Anechoic chambers duplicate the silence of space, where there is no medium (like air or water) through which the sound waves can travel. The walls of the anechoic chambers are made of carbon compound pyramids which absorb any microwave.

## Angle of Inclination

The angle at which a satellite's orbit is tilted in relation to the Earth's equator. A 90 degree angle of inclination is a polar orbit. A zero degree angle of inclination is an *equatorial orbit*.

## Anik A

A *communications satellite* launched by Telesat Canada in 1972. Anik A made Canada the first country in the world to have a satellite in geostationary orbit for domestic communications purposes. Anik A improved telephone and television communications, bringing Canadians closer together.

### Anik B

A communications satellite launched by the firm Telesat Canada on Dec 16, 1978 to replace the Anik A satellites. The purpose of the Anik satellites was to improve communications across Canada, especially for those Canadians living in remote areas.

### Anik E

A communications satellite launched by the firm Telesat Canada in 1991. Anik E was a series of two satellites—Anik E1 and Anik E2. They are the fifth generation of Anik satellites, and they are still a large part of Canadian communications.

### Anion

An ion consists of one or more atoms and carries a unit charge of electricity. Those that are negative ions (hydroxyl and acidic atoms or groups) are called anions (cf. cation).

### Anisotropic

Exhibiting different values of a property in different crystallographic directions.

### Anneal

Heat process used to remove stress, crystallize or render deposited material more uniform.

### Anode

The electrode in an electrochemical cell or galvanic couple that experiences oxidation, or gives up electrons.

### Antenna

A piece of equipment that allows transmission and reception of radio signals. Satellites need antennas to communicate with Earth. A satellite may need to receive instructions and transmit the information it collects, or it may relay the information sent to it to another site on Earth. Since the information is transmitted using radio waves, which move at the speed of light, this method allows for very fast communications (only a very small time lag).

### Aperture

A small hole. Usually with reference to a camera, the aperture is the hole in a camera that allows light to hit film. The amount of light that gets through the aperture determines what a picutre will look like.

## Apertureless NFOM (aNFOM)

This technique is based on measuring the modulation of the scattered electric field from the end of a sharp silicon tip as it is stabilized and scanned in close proximity to a sample surface. The demonstrated resolution lies in the 3 nm range—comparable to what can be achieved with typical attractive mode atomic force microscopes.

## Armored

The Molecular Plate™ Advantage Evident's Molecular Plate™ covering for InGaP EviDots is lattice matched—with a matched pattern of atoms—to improve its molecular bonding to the core. This results in a quantum dot with high brightness and increased stability for longer lasting fluorescence.

## Aromatic Hydrocarbon

An aromatic hydrocarbon (abbreviated as AH) or arene is a hydrocarbon, of which the molecular structure incorporates one or more planar sets of six carbon atoms that are connected by delocalised electrons numbering the same as if they consisted of alternating single and double covalent bonds. The term 'aromatic' was assigned before the physical mechanism determining aromaticity was discovered, and was derived from the fact that many of the compounds have a sweet scent. This sweet scent actually came from impurities in the compounds (which are not actually aromatic in the sense initially described). The configuration of six carbon atoms in aromatic compounds is known as a benzene ring, after the simplest possible such hydrocarbon, benzene. Aromatic hydrocarbons can be *monocyclic* or *polycyclic*. Some non-benzene-based compounds called heteroarenes, which follow Hückel's rule, are also aromatic compounds. In these compounds, at least one carbon atom is replaced by one of the heteroatoms oxygen, nitrogen, or sulfur. Examples of non-benzene compounds with aromatic properties are furan, a heterocyclic compound with a five-membered ring that includes an oxygen atom, and pyridine, a heterocyclic compound with a six-membered ring containing one nitrogen atom.

Benzene, $C_6H_6$, is the simplest AH and was recognized as the first aromatic hydrocarbon, with the nature of its bonding first being recognized by Friedrich August Kekulé von Stradonitz in the 19th century. Each carbon atom in the hexagonal cycle has four electrons to share. One goes to the hydrogen atom, and one each to the two neighboring carbons. This leaves one to share with one of its two neighboring carbon atoms, which is why the benzene molecule is drawn with alternating single and double bonds around the hexagon. Many chemists draw a circle around the inside of the ring to show six electrons floating around in delocalized molecular orbitals the size of the ring itself. This also accurately represents the equivalent nature of the six bonds all of bond-order ~1.5. This equivalency is well explained by resonance forms. The electrons float above and below the ring, and the electromagnetic fields they generate keep the ring flat. General properties: (i) Display aromaticity. (ii) The Carbon-Hydrogen ratio is very large. (iii) They burn with a sooty yellow flame because of the high carbon-hydrogen ratio. (iv) They undergo electrophilic substitution reactions and nucleophilic aromatic substitutions.

Benzene derivatives have from one to six substituents attached to the central benzene core. Examples of benzene compounds with just one substituent are phenol, which carries a hydroxyl group and toluene with a methyl group. When there is more than one substituent present on the ring, their spatial relationship becomes important for which the arene substitution patterns *ortho, meta,* and *para* are devised. For example, three isomers exist for cresol because the methyl group and the hydroxyl group can be placed next to each other (ortho), one position removed from each other (meta), or two positions removed from each other (para). Xylenol has two methyl groups in addition to the hydroxyl group, and, for this structure, 6 isomers exist.

## Arrhenius Equation

The equation representing the rate constant as $k = Ae^{Ea/RT}$

where A represents the product of the collision frequency and a steric factor, and $e^{Ea/RT}$ is fraction of collisions with sufficient energy to produce a reaction.

## Assembler (Molecular)

A molecular assembler as defined by K. Eric Drexler is a "proposed device able to guide chemical reactions by positioning reactive molecules with atomic precision." Some biological molecules such as ribosomes fit this definition, since while working within a cell's environment, they receive instructions from messenger RNA and then assemble specific sequences of amino acids to construct protein molecules. However, the term "molecular assembler" usually refers to theoretical human-made or synthetic devices. Development of ribosome-like molecular assemblers was funded in 2007 by the British Engineering and Physical Sciences Research Council. It is clear that molecular assemblers in this limited sense are possible. A technology roadmap project, led by the Battelle Memorial Institute and hosted by several U.S. National Laboratories has explored a range of atomically precise fabrication technologies, including both early-generation and longer-term prospects for programmable molecular assembly; the report was released in December, 2007.

However, the term "molecular assembler" has also been used in science fiction and popular culture to refer to a wide range of fantastic atom-manipulating nanomachines, many of which may be physically impossible in reality. Much of the controversy regarding "molecular assemblers" results from the confusion in the use of the name for both technical concepts and popular fantasies. In 1992, Drexler introduced the related but better-understood term "molecular manufacturing," which he defined as the programmed "chemical synthesis of complex structures by mechanically positioning reactive molecules, not by manipulating individual atoms."

Much of the body of this article discusses "molecular assemblers" in the popular sense. These include hypothetical machines that manipulate individual atoms, and machines with organism-like self-replicating abilities, mobility, ability to consume food, and so forth. These are quite different from devices that merely (as defined above) "guide chemical reactions by positioning reactive molecules with atomic precision".

Because synthetic molecular assemblers have never been constructed, and because of the confusion regarding the meaning of the term, there has been much controversy as to whether "molecular assemblers" are possible or simply science fiction. Confusion and controversy also stem from their classification as nanotechnology, which is an active area of laboratory research which has already been applied to the production of real products; however, there had been, until recently, no research efforts into the actual construction of "molecular assemblers". A primary criticism of the computational research into products of advanced "molecular assemblers" is that the structures investigated are impossible to synthesize today. A nanofactory is a proposed system in which nanomachines (resembling molecular assemblers, or industrial robot arms) would combine reactive molecules via mechanosynthesis to build larger atomically precise parts. These, in turn, would be assembled by positioning mechanisms of assorted sizes to build macroscopic (visible) but still atomically-precise products.

A typical nanofactory would fit in a desktop box, in the vision of K. Eric Drexler published in *Nanosystems: Molecular Machinery, Manufacturing and Computation* (1992), a notable work of "exploratory engineering". During the last decade, others have extended the nanofactory concept, including an analysis of nanofactory convergent assembly by Ralph Merkle, a systems design of a replicating nanofactory architecture by J. Storrs Hall, Forrest Bishop's "Universal Assembler", the patented exponential assembly process by Zyvex, and a top-level systems design for a 'primitive nanofactory' by Chris Phoenix (Director of Research at the Center for Responsible Nanotechnology). All of these nanofactory designs (and more) are summarized in Chapter 4 of *Kinematic Self-Replicating Machines* (2004) by Robert Freitas and Ralph Merkle. The Nanofactory Collaboration, founded by Robert Freitas and Ralph Merkle in 2000, is a focused ongoing effort involving 23 researchers from 10 organizations and 4 countries that is developing a practical research agenda specifically aimed at positionally-controlled diamond mechanosynthesis and diamondoid nanofactory development. In 2005, a computer-animated short film of the nanofactory concept was produced by John Burch, in collaboration with Drexler. Such visions have been the subject of much debate, on several intellectual levels. No one has discovered an insurmountable problem with the underlying theories and no one has proved that the theories can be translated into practice. However, the debate continues, with some of it being summarized in the Molecular nanotechnology article.

If nanofactories could be built, severe disruption to the world economy would be one of many possible negative impacts. Great benefits also would be anticipated. Various works of science fiction have explored these and similar concepts. The potential for such devices was part of the mandate of a major UK study led by mechanical engineering professor Dame Ann Dowling. The report is now complete.

*Self-replication:* "Molecular assemblers" have been confused with self-replicating machines. The nanoscale size of a typical science fiction universal molecular assembler requires an extremely large number of such devices in order to produce a practical quantity of a desired product. However, if one were able to construct a single such molecular assembler then it might be programmed to self-replicate, constructing many copies of itself, allowing an exponential rate of production. Then after sufficient quantities of the molecular assemblers were available, they would then be re-programmed for production of the desired product. However, if self-replication of molecular assemblers were not restrained then it might lead to competition with naturally occurring organisms. This has been called ecophagy or the grey goo problem. One method to building molecular assemblers is to mimic evolutionary processes employed by biological systems. Biological evolution proceeds by random variation combined with culling of the less-successful variants and reproduction of the more-successful variants. Production of complex molecular assemblers might be evolved from simpler systems since "A complex system that works is invariably found to have evolved from a simple system that worked. . . . A complex system designed from scratch never works and can not be patched up to make it work. You have to start over, beginning with a system that works." However, most published safety guidelines include "recommendations against developing ... replicator designs which permit surviving mutation or undergoing evolution".

Most assembler designs keep the "source code" external to the physical assembler. At each step of a manufacturing process, that step is read from an ordinary computer file and "broadcast" to all the assemblers. If any assembler gets out of range of that computer, or when the link between that computer and the assemblers is broken, or when that computer is unplugged, the assemblers stop replicating. Such a "broadcast architecture" is one of the safety features recommended by the "Foresight Guidelines on Molecular Nanotechnology", and a map of the 137-dimensional replicator design space recently published by Freitas and Merkle provides numerous practical methods by which replicators can be safely controlled by good design.

*Drexler and Smalley debate:* One of the most outspoken critics of some concepts of "molecular assemblers" was Professor Richard Smalley (1943-2005) who won the Nobel prize for his contributions to the field of nanotechnology. Smalley believed that such assemblers were not physically possible and introduced scientific objections to them. His two principal technical objections were termed the "fat fingers problem" and the "sticky fingers problem" that he believed would exclude the possibility of "molecular assemblers" that worked by precision picking and placing of individual atoms. Drexler and coworkers have responded to these two issues in a 2001 publication.

Smalley also believed that Drexler's speculations about apocalyptic dangers of self-replicating machines that have been equated with "molecular assemblers" would threaten the public support for development of nanotechnology. To address the debate between Drexler and Smalley regarding molecular assemblers *Chemical & Engineering News* published a point-counterpoint consisting of an exchange of letters that addressed the issues.

*Regulation:* Speculation on the power of systems that have been called "molecular assemblers" has sparked a wider political discussion on the implication of nanotechnology. This is in part due to the fact that nanotechnology is a very broad term and could include "molecular assemblers." Discussion of the possible implications of fantastic molecular assemblers has prompted calls for regulation of current and future nanotechnology. There are very real concerns with the potential health and ecological impact of nanotechnology that is being integrated in manufactured products. Greenpeace for instance commissioned a report concerning nanotechnology in which they express concern into the toxicity of nanomaterials that have been introduced in the environment. However, it makes only passing references to "assembler" technology. The UK Royal Society and UK Royal Academy of Engineering also commissioned a report entitled "Nanoscience and nanotechnologies: opportunities and uncertainties" regarding the larger social and ecological implications on nanotechnology. This report does not discuss the threat posed by potential so-called "molecular assemblers."

*Grey goo:* Speculation about the potential dangers of artificial self replicating machines (which have been labeled as "molecular assemblers") has led some to envision apocalyptic scenarios. One scenario suggested danger to life could arise in the form of grey goo which consumes carbon to make more of itself. If unchecked such mechanical replication could potentially consume whole ecoregions or the whole Earth (ecophagy), or it could simply outcompete other natural lifeforms for necessary resources such as carbon, ATP, or UV light (which some nanomotor examples run on). It is worth noting that the ecophagy and 'grey goo' scenarios, like synthetic molecular assemblers, are based upon still-theoretical technologies that have not yet been demonstrated experimentally.

## Astronomy Satellite

An astronomy satellite's vision is not clouded by the gases that make up the Earth's atmosphere, unlike that of telescopes on Earth. Astronomy satellites study stellar phenomenona like black holes, quasars, and distant galaxies. These are not to be confused with space exploration satellites, which also study these phenomena.

## At the Core

Non-Heavy Metal Quantum Dot. Research at Evident, which is focused on advancing product development, led to the first commercially available nanocrystals that comprise a III-V semiconductor material: Indium, Gallium and Phosphide core quantum dots, with a unique molecular metallic platting compound to form the shell or InGaP/ZnS EviDots-MP™.

## Atmospheric Studies Satellite

A type of scientific satellite that studies the Earth's atmosphere. They were some of the very first satellites launched into space, including Canada's first satellite Alouette.

## Atom

The atom is the smallest unit of an element that retains the chemical properties of that element. An atom has an electron cloud consisting of negatively charged electrons surrounding a dense nucleus. The nucleus contains positively charged protons and electrically neutral neutrons. When the number of protons in the nucleus equals the number of electrons, the atom is electrically neutral; otherwise it is an ion and has a net positive or negative charge. An atom is classified according to its number of protons and neutrons: the number of protons determines the chemical element and the number of neutrons determines the isotope of that element. Something that cannot be divided further. The concept of an atom as an indivisible component of matter was first proposed by early Indian and Greek philosophers. In the 17th and 18th centuries, chemists provided a physical basis for this idea by showing that certain substances could not be further broken down by chemical methods. During the late 19th and early 20th centuries, physicists discovered subatomic components and structure inside the atom, thereby demonstrating that the 'atom' was not indivisible. The principles of quantum mechanics were used to successfully model the atom.

Relative to everyday experience, atoms are minuscule objects with proportionately tiny masses that can only be observed individually using special instruments such as the scanning tunneling microscope. Over 99.9% of an atom's mass is concentrated in the nucleus, with protons and neutrons having roughly equal mass. Each element has at least one isotope with unstable nuclei that can undergo radioactive decay. This can result in a transmutation that changes the number of protons or neutrons in a nucleus. Electrons occupy a set of stable energy levels, or orbitals, and can transition between these states by absorbing or emitting photons that match the energy differences between the levels. The electrons determine the chemical properties of an element, and strongly influence an atom's magnetic properties. The concept that matter is composed of discrete units and cannot be divided into arbitrarily tiny quantities has been around for millennia, but these ideas were founded in abstract, philosophical reasoning rather than experimentation and empirical observation. The nature of atoms in philosophy varied considerably over time and between cultures and schools, and often had spiritual elements. Nevertheless, the basic idea of the atom was adopted by scientists thousands of years later because it elegantly explained new discoveries in the field of chemistry.

The earliest references to the concept of atoms date back to ancient India in the 6th century BCE. The Nyaya and Vaisheshika schools developed elaborate theories of how atoms combined into more complex objects (first in pairs, then trios of pairs). The references to atoms in the West emerged a century later from Leucippus whose student, Democritus, systemized his views. In approximately 450 BCE, Democritus coined the term *átomos* (Greek: ôïìïò), which means "uncuttable" or "the smallest indivisible particle of matter", i.e., something that cannot be divided. Although the Indian and Greek concepts of the atom were based purely on philosophy, modern science has retained the name coined by Democritus. Further progress in the understanding of atoms did not occur until the science of chemistry began to develop. In 1661, natural philosopher Robert Boyle published *The Sceptical Chymist* in which he argued that matter was composed of various combinations of different "corpuscules" or atoms, rather than the classical elements of air, earth, fire and water. In 1789 the term *element* was defined by the French nobleman and scientific researcher Antoine Lavoisier to mean basic substances that could not be further broken down by the methods of chemistry.

In 1803, English instructor and natural philosopher John Dalton used the concept of atoms to explain why elements always react in a ratio of small whole numbers—the law of multiple proportions—and why certain gases dissolve better in water than others. He proposed that each element consists of atoms of a single, unique type, and that these atoms can join together to form chemical compounds.

Additional validation of particle theory (and by extension atomic theory) occurred in 1827 when botanist Robert Brown used a microscope to look at dust grains floating in water and discovered that they moved about erratically—a phenomenon that became known as "Brownian motion". J. Desaulx suggested in 1877 that the phenomenon was caused by the thermal motion of water molecules, and in 1905 Albert Einstein produced the first mathematical analysis of the motion, thus confirming the hypothesis.

The physicist J. J. Thomson, through his work on cathode rays in 1897, discovered the electron and its subatomic nature, which destroyed the concept of atoms as being indivisible units. Thomson believed that the electrons were distributed throughout the atom, with their charge balanced by the presence of a uniform sea of positive charge (the plum pudding model). However, in 1909, researchers under the direction of physicist Ernest Rutherford bombarded a sheet of gold foil with helium ions and discovered that a small percentage were deflected through much larger angles than was predicted using Thomson's proposal. Rutherford interpreted the gold foil experiment as suggesting that the positive charge of an atom and most of its mass was concentrated in a nucleus at the center of the atom (the Rutherford model), with the electrons orbiting it like planets around a sun. Positively charged helium ions passing close to this dense nucleus would then be deflected away at much sharper angles.

In 1803, English instructor and natural philosopher John Dalton used the concept of atoms to explain why elements always react in a ratio of small whole numbers—the law of multiple proportions—and why certain gases dissolve better in water than others. He proposed that each element consists of atoms of a single, unique type, and that these atoms can join together to form chemical compounds. Additional validation of particle theory (and by extension atomic theory) occurred in 1827 when botanist Robert Brown used a microscope to look at dust grains floating in water and discovered that they moved about erratically—a phenomenon that became known as "Brownian motion". J. Desaulx suggested in 1877 that the phenomenon was caused by the thermal motion of water molecules, and in 1905 Albert Einstein produced the first mathematical analysis of the motion, thus confirming the hypothesis.

The physicist J. J. Thomson, through his work on cathode rays in 1897, discovered the electron and its subatomic nature, which destroyed the concept of atoms as being indivisible units. Thomson believed that the electrons were distributed throughout the atom, with their charge balanced by the presence of a uniform sea of positive charge (the plum pudding model).

However, in 1909, researchers under the direction of physicist Ernest Rutherford bombarded a sheet of gold foil with helium ions and discovered that a small percentage were deflected through much larger angles than was predicted using Thomson's proposal. Rutherford interpreted the gold foil experiment as suggesting that the positive charge of an atom and most of its mass was concentrated in a nucleus at the center of the atom (the Rutherford model), with the electrons orbiting it like planets around a sun. Positively charged helium ions passing close to this dense nucleus would then be deflected away at much sharper angles.

Meanwhile, in 1913, physicist Niels Bohr revised Rutherford's model by suggesting that the electrons were confined into clearly defined orbits, and could jump between these, but could not freely spiral inward or outward in intermediate states. An electron must absorb or emit specific amounts of energy to transition between these fixed orbits. When the light from a heated material is passed through a prism, it produced a multi-colored spectrum. The appearance of fixed lines in this spectrum was successfully explained by the orbital transitions.

Chemical bonds between atoms were now explained, by Gilbert Newton Lewis in 1916, as the interactions between their constituent electrons. As the chemical properties of the elements were known to largely repeat themselves according to the periodic law, in 1919 the American chemist Irving Langmuir suggested that this could be explained if the electrons in an atom were connected or clustered in some manner. Groups of electrons were thought to occupy a set of electron shells about the nucleus.

In 1926, Erwin Schrödinger, using Louis de Broglie's 1924 proposal that particles behave to an extent like waves, developed a mathematical model of the atom that described the electrons as three-dimensional waveforms, rather than point particles. A consequence of using waveforms to describe electrons is that it is mathematically impossible to obtain precise values for both the position and momentum of a particle at the same time; this became known as the uncertainty principle, formulated by Werner Heisenberg in 1926. In this concept, for each measurement of a position one could only obtain a range of probable values for momentum, and vice versa. Although this model was difficult to visualise, it was able to explain observations of atomic behavior that previous models could not, such as certain structural and spectral patterns of atoms larger than hydrogen. Thus, the planetary model of the atom was discarded in favor of one that described atomic orbital zones around the nucleus where a given electron is most likely to exist.

The development of the mass spectrometer allowed the exact mass of atoms to be measured. The device uses a magnet to bend the trajectory of a beam of ions, and the amount of deflection is determined by the ratio of an atom's mass to its charge. The chemist Francis William Aston used this instrument to demonstrate that isotopes had different masses. The mass of these isotopes varied by integer amounts, called the whole number rule. The explanation for these different atomic isotopes awaited the discovery of the neutron, a neutral-charged particle with a mass similar to the proton, by the physicist James Chadwick in 1932. Isotopes were then explained as elements with the same number of protons, but different numbers of neutrons within the nucleus.

In the 1950s, the development of improved particle accelerators and particle detectors allowed scientists to study the impacts of atoms moving at high energies. Neutrons and protons were found to be hadrons, or composites of smaller particles called quarks. Standard models of nuclear physics were developed that successfully explained the properties of the nucleus in terms of these sub-atomic particles and the forces that govern their interactions.

Around 1985, Steven Chu and co-workers at Bell Labs developed a technique for lowering the temperatures of atoms using lasers. In the same year, a team led by William D. Phillips managed to contain atoms of sodium in a magnetic trap. The combination of these two techniques and a method based on the Doppler effect, developed by Claude Cohen-Tannoudji and his group, allows small numbers of atoms to be cooled to several microkelvin. This allows the atoms to be studied with great precision, and later led to the discovery of Bose-Einstein condenzation. Historically, single atoms have been prohibitively small for scientific applications. Recently, devices have been constructed that use a single metal atom connected through organic ligands to construct a single electron transistor. Experiments have been carried out by trapping and slowing single atoms using laser cooling in a cavity to gain a better physical understanding of matter.

Though the word *atom* originally denoted a particle that cannot be cut into smaller particles, in modern scientific usage the atom is composed of various subatomic particles. The constituent particles of an atom are the electron, the proton and the neutron, except that hydrogen-1 has no neutrons and a positive hydrogen ion has no electrons. The electron is by far the least massive of these particles at 9.11 × 10 kg, with a negative electrical charge and a size that is too small to be measured using available techniques. Protons have a positive charge and a mass 1,836 times that of the electron, at 1.6726 × 10 kg, although this can be reduced by changes to the atomic binding energy. Neutrons have no electrical charge and have a free mass of 1,839 times the mass of electrons, or 1.6929 × 10 kg. Neutrons and protons have comparable dimensions—on the order of 2.5 × 10 m—although the 'surface' of these particles is not sharply defined.

In the Standard Model of physics, both protons and neutrons are composed of elementary particles called quarks. The quark is a type of fermion, and is one of the two basic constituents of matter—the other being the lepton, of which the electron is an example. There are six types of quarks, each having a fractional electric charge of either +2/3 or "1/3. Protons are composed of two up quarks and one down quark, while a neutron consists of one up quark and two down quarks. This distinction accounts for the difference in mass and charge between the two particles. The quarks are held together by the strong nuclear force, which is mediated by gluons. The gluon is a member of the family of gauge bosons, which are elementary particles that mediate physical forces.

All the bound protons and neutrons in an atom make up a tiny atomic nucleus, and are collectively called nucleons. The radius of a nucleus is approximately equal to  fm, where *A* is the total number of nucleons. This is much smaller than the radius of the atom, which is on the order of 10 fm. The nucleons are bound together by a short-ranged attractive potential called the residual strong force. At distances smaller than 2.5 fm this force is much more powerful than the electrostatic force that causes positively charged protons to repel each other. Atoms of the same element have the same number of protons, called the atomic number. Within a single element, the number of neutrons may vary, determining the isotope of that element. The total number of protons and neutrons determine the nuclide. The number of neutrons relative to the protons determines the stability of the nucleus, with certain isotopes undergoing radioactive decay.

The neutron and the proton are different types of fermions. The Pauli exclusion principle is a quantum mechanical effect that prohibits *identical* fermions (such as multiple protons) from occupying the same quantum physical state at the same time. Thus every proton in the nucleus must occupy a different state, with its own energy level, and the same rule applies to all of the neutrons. (This prohibition does not apply to a proton and neutron occupying the same quantum state.) A nucleus that has a different number of protons than neutrons can potentially drop to a lower energy state through a radioactive decay that causes the number of protons and neutrons to more closely match. As a result, atoms with roughly matching numbers of protons and neutrons are more stable against decay. However, with increasing atomic number, the mutual repulsion of the protons requires an increasing proportion of neutrons to maintain the stability of the nucleus, which modifies this trend. Thus, there are no stable nuclei with equal proton and neutron numbers above atomic number Z = 20 (calcium); and as Z increases toward the heaviest nuclei, the ratio of neutrons per proton required for stability increases to about 1.5.

The number of protons and neutrons in the atomic nucleus can be modified, although this can require very high energies because of the strong force. Nuclear fusion occurs when multiple atomic particles join to form a heavier nucleus, such as through the energetic collision of two nuclei. At the core of the Sun, protons require energies of 3–10 KeV to overcome their mutual repulsion—the coulomb barrier—and fuse together into a single nucleus. Nuclear fission is the opposite process, causing a nucleus to split into two smaller nuclei—usually through radioactive decay. The nucleus can also be modified through bombardment by high energy subatomic particles or photons. In such processes that change the number of protons in a nucleus, the atom becomes an atom of a different chemical element.

If the mass of the nucleus following a fusion reaction is less than the sum of the masses of the separate particles, then the difference between these two values is emitted as energy, as described by Albert Einstein's mass–energy equivalence formula, $E = mc$, where $m$ is the mass loss and $c$ is the speed of light. This deficit is the binding energy of the nucleus. The fusion of two nuclei that have lower atomic numbers than iron and nickel is usually an exothermic process that releases more energy than is required to bring them together. It is this energy-releasing process that makes nuclear fusion in stars a self-sustaining reaction. For heavier nuclei, the total binding energy begins to decrease. That means fusion processes with nuclei that have higher atomic numbers is an endothermic process. These more massive nuclei can not undergo an energy-producing fusion reaction that can sustain the hydrostatic equilibrium of a star.

The electrons in an atom are attracted to the protons in the nucleus by the electromagnetic force. This force binds the electrons inside an electrostatic potential well surrounding the smaller nucleus, which means that an external source of energy is needed in order for the electron to escape. The closer an electron is to the nucleus, the greater the attractive force. Hence electrons bound near the center of the potential well require more energy to escape than those at the exterior. Electrons, like other particles, have properties of both a particle and a wave. The electron cloud is a region inside the potential well where each electron forms a type of three-dimensional standing wave—a wave form that does not move relative to the nucleus. This behavior is defined by an atomic orbital, a mathematical function that characterises the probability that an electron will appear to be at a particular location when its position is measured. Only a discrete (or quantized) set of these orbitals exist around the nucleus, as other possible wave patterns will rapidly decay into a more stable form. Orbitals can have one or more ring or node structures, and they differ from each other in size, shape and orientation.

Each atomic orbital corresponds to a particular energy level of the electron. The electron can change its state to a higher energy level by absorbing a photon with sufficient energy to boost it into the new quantum state. Likewise, through spontaneous emission, an electron in a higher energy state can drop to a lower energy state while radiating the excess energy as a photon. These characteristic energy values, defined by the differences in the energies of the quantum states, are responsible for atomic spectral lines. The amount of energy needed to remove or add an electron (the electron binding energy) is far less than the binding energy of nucleons. For example, it requires only 13.6 eV to strip a ground-state electron from a hydrogen atom. Atoms are electrically neutral if they have an equal number of protons and electrons. Atoms that have either a deficit or a surplus of electrons are called ions. Electrons that are farthest from the nucleus may be transferred to other nearby atoms or shared between atoms. By this mechanism, atoms are able to bond into molecules and other types of chemical compounds like ionic and covalent network crystals.

By definition, any two atoms with an identical number of *protons* in their nuclei belong to the same chemical element. Atoms with equal numbers of protons but a different number of *neutrons* are different isotopes of the same element. For example, all hydrogen atoms admit exactly one proton, but isotopes exist with no neutrons (hydrogen-1, by far the most common form, sometimes called protium), one neutron (deuterium), two neutrons (tritium) and more than two neutrons. The known elements form a set of atomic numbers from hydrogen with a single proton up to the 118-proton element ununoctium. All known isotopes of elements with atomic numbers greater than 82 are radioactive. About 339 nuclides occur naturally on Earth, of which 269 (about 79%) are stable. Of the chemical elements, 80 have one or more stable isotopes. Elements 43, 61, and all elements numbered 83 or higher have no stable isotopes. As a rule, there is, for each atomic number (each element) only a handful of stable isotopes, the average being 3.4 stable isotopes per element which has any stable isotopes. Sixteen elements have only a single stable isotope, while the largest number of stable isotopes observed for any element is ten (for the element tin).

Stability of isotopes is affected by the ratio of protons to neutrons, and also by presence of certain "magic numbers" of neutrons or protons which represent closed and filled quantum shells. These quantum shells correspond to a set of energy levels within the shell model of the nucleus. Of the 269 known stable nuclides, only four have both an odd number of protons *and* odd number of neutrons: H, Li, B and N. Also, only four naturally-occurring, radioactive odd-odd nuclides have a half-life over a billion years: K, V, La and Ta. Most odd-odd nuclei are highly unstable with respect to beta decay, because the decay products are even-even, and are therefore more strongly bound, due to nuclear pairing effects.

Because the large majority of an atom's mass comes from the protons and neutrons, the total number of these particles in an atom is called the mass number. The mass of an atom at rest is often expressed using the unified atomic mass unit (u), which is also called a Dalton (Da). This unit is defined as a twelfth of the mass of a free neutral atom of carbon-12, which is approximately 1.66 × 10 kg. Hydrogen-1, the lightest isotope of hydrogen and the atom with the lowest mass, has an atomic weight of 1.007825 u. An atom has a mass approximately equal to the mass number times the atomic mass unit. The heaviest stable atom is lead-208, with a mass of 207.9766521 u. As even the most massive atoms are far too light to work with directly, chemists instead use the unit of moles. The mole is defined such that one mole of any element will always have the same number of atoms (about 6.022 × 10). This number was chosen so that if an element has an atomic mass of 1 u, a mole of atoms of that element will have a mass of 0.001 kg, or 1 gram. Carbon, for example, has an atomic mass of 12 u, so a mole of carbon atoms weighs 0.012 kg.

Atoms lack a well-defined outer boundary, so the dimensions are usually described in terms of the distances between two nuclei when the two atoms are joined in a chemical bond. The radius varies with the location of an atom on the atomic chart, the type of chemical bond, the number of neighboring atoms (coordination number) and a quantum mechanical property known as spin. On the periodic table of the elements, atom size tends to increase when moving down columns, but decrease when moving across rows (left to right). Consequently, the smallest atom is helium with a radius of 32 pm, while one of the largest is caesium at 225 pm. These dimensions are thousands of times smaller than the wavelengths of light (400–700 nm) so they can not be viewed using an optical microscope. However, individual atoms can be observed using a scanning tunneling microscope. Some examples will demonstrate the minuteness of the atom. A typical human hair is about 1 million carbon atoms in width. A single drop of water contains about 2 sextillion (2 × 10) atoms of oxygen, and twice the number of hydrogen atoms. A single carat diamond with a mass of 2 × 10 kg contains about 10 sextillion atoms of carbon. If an apple was magnified to the size of the Earth, then the atoms in the apple would be approximately the size of the original apple. Every element has one or more isotopes that have unstable nuclei that are subject to radioactive decay, causing the nucleus to emit particles or electromagnetic radiation. Radioactivity can occur when the radius of a nucleus is large compared with the radius of the strong force, which only acts over distances on the order of 1 fm.

The most common forms of radioactive decay are: (i) Alpha decay is caused when the nucleus emits an alpha particle, which is a helium nucleus consisting of two protons and two neutrons. The result of the emission is a new element with a lower atomic number. (ii) Beta decay is regulated by the weak force, and results from a transformation of a neutron into a proton, or a proton into a neutron. The first is accompanied by the emission of an electron and an antineutrino, while the second causes the emission of a positron and a neutrino. The electron or positron emissions are called beta particles. Beta decay either increases or decreases the atomic number of the nucleus by one. (iii) Gamma decay results from a change in the energy level of the nucleus to a lower state, resulting in the emission of electromagnetic radiation. This can occur following the emission of an alpha or a beta particle from radioactive decay.

Other more rare types of radioactive decay include ejection of neutrons or protons or clusters of nucleons from a nucleus, or more than one beta particle, or result (through internal conversion) in production of high-speed electrons which are not beta rays, and high-energy photons which are not gamma rays. Each radioactive isotope has a characteristic decay time period—the half-life—that is determined by the amount of time needed for half of a sample to decay. This is an exponential decay process that steadily decreases the proportion of the remaining isotope by 50% every half life. Hence after two half-lives have passed only 25% of the isotope will be present, and so forth.

Elementary particles possess an intrinsic quantum mechanical property known as spin. This is analogous to the angular momentum of an object that is spinning around its center of mass, although strictly speaking these particles are believed to be point-like and cannot be said to be rotating. Spin is measured in units of the reduced Planck constant (), with electrons, protons and neutrons all having spin ½ , or "spin-½". In an atom, electrons in motion around the nucleus possess orbital angular momentum in addition to their spin, while the nucleus itself possesses angular momentum due to its nuclear spin. The magnetic field produced by an atom—its magnetic moment—is determined by these various forms of angular momentum, just as a rotating charged object classically produces a magnetic field. However, the most dominant contribution comes from spin. Due to the nature of electrons to obey the Pauli exclusion principle, in which no two electrons may be found in the same quantum state, bound electrons pair up with each other, with one member of each pair in a spin up state and the other in the opposite, spin down state. Thus these spins cancel each other out, reducing the total magnetic dipole moment to zero in some atoms with even number of electrons.

In ferromagnetic elements such as iron, an odd number of electrons leads to an unpaired electron and a net overall magnetic moment. The orbitals of neighboring atoms overlap and a lower energy state is achieved when the spins of unpaired electrons are aligned with each other, a process known as an exchange interaction. When the magnetic moments of ferromagnetic atoms are lined up, the material can produce a measurable macroscopic field. Paramagnetic materials have atoms with magnetic moments that line up in random directions when no magnetic field is present, but the magnetic moments of the individual atoms line up in the presence of a field.

The nucleus of an atom can also have a net spin. Normally these nuclei are aligned in random directions because of thermal equilibrium. However, for certain elements (such as xenon-129) it is possible to polarize a significant proportion of the nuclear spin states so that they are aligned in the same direction—a condition called hyperpolarization. This has important applications in magnetic resonance imaging. When an electron is bound to an atom, it has a potential energy that is inversely proportional to its distance from the nucleus. This is measured by the amount of energy needed to unbind the electron from the atom, and is usually given in units of electronvolts (eV). In the quantum mechanical model, a bound electron can only occupy a set of states centered on the nucleus, and each state corresponds to a specific energy level. The lowest energy state of a bound electron is called the ground state, while an electron at a higher energy level is in an excited state. In order for an electron to transition between two different states, it must absorb or emit a photon at an energy matching the difference in the potential energy of those levels. The energy of an emitted photon is proportional to its frequency, so these specific energy levels appear as distinct bands in the electromagnetic spectrum. Each element has a characteristic spectrum that can depend on the nuclear charge, subshells filled by electrons, the electromagnetic interactions between the electrons and other factors.

When a continuous spectrum of energy is passed through a gas or plasma, some of the photons are absorbed by atoms, causing electrons to change their energy level. Those excited electrons that remain bound to their atom will spontaneously emit this energy as a photon, traveling in a random direction, and so drop back to lower energy levels. Thus the atoms behave like a filter that forms a series of dark absorption bands in the energy output. (An observer viewing the atoms from a different direction, which does not include the continuous spectrum in the background, will instead see a series of emission lines from the photons emitted by the atoms.) Spectroscopic measurements of the strength and width of spectral lines allow the composition and physical properties of a substance to be determined.

Close examination of the spectral lines reveals that some display a fine structure splitting. This occurs because of spin-orbit coupling, which is an interaction between the spin and motion of the outermost electron. When an atom is in an external magnetic field, spectral lines become split into three or more components; a phenomenon called the Zeeman effect. This is caused by the interaction of the magnetic field with the magnetic moment of the atom and its electrons. Some atoms can have multiple electron configurations with the same energy level, which thus appear as a single spectral line. The interaction of the magnetic field with the atom shifts these electron configurations to slightly different energy levels, resulting in multiple spectral lines. The presence of an external electric field can cause a comparable splitting and shifting of spectral lines by modifying the electron energy levels, a phenomenon called the Stark effect. If a bound electron is in an excited state, an interacting photon with the proper energy can cause stimulated emission of a photon with a matching energy level. For this to occur, the electron must drop to a lower energy state that has an energy difference matching the energy of the interacting photon. The emitted photon and the interacting photon will then move off in parallel and with matching phases. That is, the wave patterns of the two photons will be synchronized. This physical property is used to make lasers, which can emit a coherent beam of light energy in a narrow frequency band.

The outermost electron shell of an atom in its uncombined state is known as the valence shell, and the electrons in that shell are called valence electrons. The number of valence electrons determines the bonding behavior with other atoms. Atoms tend to chemically react with each other in a manner that will fill (or empty) their outer valence shells. The chemical elements are often displayed in a periodic table that is laid out to display recurring chemical properties, and elements with the same number of valence electrons form a group that is aligned in the same column of the table. (The horizontal rows correspond to the filling of a quantum shell of electrons.) The elements at the far right of the table have their outer shell completely filled with electrons, which results in chemically inert elements known as the noble gases.

Quantities of atoms are found in different states of matter that depend on the physical conditions, such as temperature and pressure. By varying the conditions, materials can transition between solids, liquids, gases and plasmas. Within a state, a material can also exist in different phases. An example of this is solid carbon, which can exist as graphite or diamond. At temperatures close to absolute zero, atoms can form a Bose–Einstein condensate, at which point quantum mechanical effects, which are normally only observed at the atomic scale, become apparent on a macroscopic scale. This supercooled collection of atoms then behaves as a single super atom, which may allow fundamental checks of quantum mechanical behavior.

The scanning tunneling microscope is a device for viewing surfaces at the atomic level. It uses the quantum tunneling phenomenon, which allows particles to pass through a barrier that would normally be insurmountable. Electrons tunnel through the vacuum between two planar metal electrodes, on each of which is an adsorbed atom, providing a tunneling-current density that can be measured. Scanning one atom (taken as the tip) as it moves past the other (the sample) permits plotting of tip displacement versus lateral separation for a constant current. The calculation shows the extent to which scanning-tunneling-microscope images of an individual atom are visible. It confirms that for low bias, the microscope images the space-averaged dimensions of the electron orbitals across closely packed energy levels—the

Fermi level local density of states. An atom can be ionized by removing one of its electrons. The electric charge causes the trajectory of an atom to bend when it passes through a magnetic field. The radius by which the trajectory of a moving ion is turned by the magnetic field is determined by the mass of the atom. The mass spectrometer uses this principle to measure the mass-to-charge ratio of ions. If a sample contains multiple isotopes, the mass spectrometer can determine the proportion of each isotope in the sample by measuring the intensity of the different beams of ions. Techniques to vaporize atoms include inductively coupled plasma atomic emission spectroscopy and inductively coupled plasma mass spectrometry, both of which use a plasma to vaporize samples for analysis. A more area-selective method is electron energy loss spectroscopy, which measures the energy loss of an electron beam within a transmission electron microscope when it interacts with a portion of a sample. The atom-probe tomograph has sub-nanometer resolution in 3-D and can chemically identify individual atoms using time-of-flight mass spectrometry.

Spectra of excited states can be used to analyze the atomic composition of distant stars. Specific light wavelengths contained in the observed light from stars can be separated out and related to the quantized transitions in free gas atoms. These colors can be replicated using a gas-discharge lamp containing the same element. Helium was discovered in this way in the spectrum of the Sun 23 years before it was found on Earth.

Atoms form about 4% of the total mass density of the observable universe, with an average density of about 0.25 atoms/m. Within a galaxy such as the Milky Way, atoms have a much higher concentration, with the density of matter in the interstellar medium (ISM) ranging from 10 to 10 atoms/m. The Sun is believed to be inside the Local Bubble, a region of highly ionized gas, so the density in the solar neighborhood is only about 10 atoms/m. Stars form from dense clouds in the ISM, and the evolutionary processes of stars result in the steady enrichment of the ISM with elements more massive than hydrogen and helium. Up to 95% of the Milky Way's atoms are concentrated inside stars and the total mass of atoms forms about 10% of the mass of the galaxy. (The remainder of the mass is an unknown dark matter.)

Stable protons and electrons appeared one second after the Big Bang. During the following three minutes, Big Bang nucleosynthesis produced most of the helium, lithium, and deuterium in the universe, and perhaps some of the beryllium and boron. The first atoms (complete with bound electrons) were theoretically created 380,000 years after the Big Bang—an epoch called recombination, when the expanding universe cooled enough to allow electrons to become attached to nuclei. Since then, atomic nuclei have been combined in stars through the process of nuclear fusion to produce elements up to iron. Isotopes such as lithium-6 are generated in space through cosmic ray spallation. This occurs when a high-energy proton strikes an atomic nucleus, causing large numbers of nucleons to be ejected. Elements heavier than iron were produced in supernovae through the r-process and in AGB stars through the s-process, both of which involve the capture of neutrons by atomic nuclei. Elements such as lead formed largely through the radioactive decay of heavier elements.

Most of the atoms that make up the Earth and its inhabitants were present in their current form in the nebula that collapsed out of a molecular cloud to form the solar system. The rest are the result of radioactive decay, and their relative proportion can be used to determine the age of the Earth through radiometric dating. Most of the helium in the crust of the Earth (about 99% of the helium from gas wells, as shown by its lower abundance of helium-3) is a product of alpha decay. There are a few trace atoms on Earth that were not present at the beginning (i.e., not "primordial"), nor are results of radioactive decay. Carbon-14 is continuously generated by cosmic rays in the atmosphere. Some atoms on Earth have been artificially generated either deliberately or as by-products of nuclear reactors or explosions. Of the transuranic elements—those with atomic numbers greater than 92—only plutonium and neptunium occur naturally on Earth. Transuranic elements have radioactive lifetimes shorter than the current age of the Earth and thus identifiable quantities of these elements have long since decayed, with the exception of traces of plutonium-244 possibly deposited by cosmic dust. Natural deposits of plutonium and neptunium are produced by neutron capture in uranium ore.

The Earth contains approximately 1.33 × 10 atoms. In the planet's atmosphere, small numbers of independent atoms of noble gases exist, such as argon and neon. The remaining 99% of the atmosphere is bound in the form of molecules, including carbon dioxide and diatomic oxygen and nitrogen. At the surface of the Earth, atoms combine to form various compounds, including water, salt, silicates and oxides. Atoms can also combine to create materials that do not consist of discrete molecules, including crystals and liquid or solid metals. This atomic matter forms networked arrangements that lack the particular type of small-scale interrupted order associated with molecular matter.

While isotopes with atomic numbers higher than lead (82) are known to be radioactive, an "island of stability" has been proposed for some elements with atomic numbers above 103. These superheavy elements may have a nucleus that is relatively stable against radioactive decay. The most likely candidate for a stable superheavy atom, unbihexium, has 126 protons and 184 neutrons. Each particle of matter has a corresponding antimatter particle with the opposite electrical charge. Thus, the positron is a positively charged antielectron and the antiproton is a negatively charged equivalent of a proton. For unknown reasons, antimatter particles are rare in the universe, hence, no antimatter atoms have been discovered in nature. However, antihydrogen, the antimatter counterpart of hydrogen, was first synthesized at the CERN laboratory in Geneva in 1996. Other exotic atoms have been created by replacing one of the protons, neutrons or electrons with other particles that have the same charge. For example, an electron can be replaced by a more massive muon, forming a muonic atom. These types of atoms can be used to test the fundamental predictions of physics.

## Atom Probe

The atom probe made one-dimensional compositional maps by combining time-of-flight spectroscopy and field ion microscopy (FIM). The instrument allows the three-dimensional reconstruction of up-to hundreds-of-millions of atoms from a sharp tip (corresponding to specimen volumes of 10,000-1,000,000 $nm^3$).

As in FIM, a sharp tip is made, placed in ultra high vacuum at cryogenic temperature (typically 20-100 K). A region of the tip's surface is selected (sometimes from an FIM image) and placed over a "probe hole" by moving the tip. The atoms at the apex of the tip are ionized, either by a positive pulsed voltage or a laser. These ions are repelled from the tip electrostatically and those passing through the probe hole reach a detector. A fast timing circuit is used to measure the time taken between the pulse and the impact of the ion on the detector, thus allowing the mass-to-charge ratio of the ion to be calculated and; therefore, the corresponding element (or elements) to be identified. From the collection of many of these ions, a chemical profile of the sample can be made with relative position accuracy of less than one atomic spacing.

The imaging atom probe (IAP), invented in 1974 by J. A. Panitz, decreased the need to moving the tip. In the IAP, ions emitted from the surface are recorded and mass analyzed at a detector placed within 12 cm of the tip (to provide a reasonably large field of view). By "time-gating" the detector for the arrival of a particular species of interest its crystallographic distribution on the surface, and as a function of depth, can be determined. Without time-gating all of the species reaching the detector are analyzed

Atom-probe tomography (APT) uses a position-sensitive detector to deduce the lateral location of atoms. This allows 3-D reconstructions to be generated. The idea of the APT, inspired by J. A. Panitz's patent, was developed by Mike Miller starting in 1983 and culminated with the first prototype in 1986. Various refinements were made to the instrument, including the use of a so-called position-sensitive (PoS) detector by Alfred Cerezo, Terence Godfrey, and George D. W. Smith in 1988. This PoSAP was commercialized by the developers. Since then, there have been many refinements to increase the field of view, mass and position resolution, and data acquisition rate of the instrument. Imago Scientific Instruments (Madison, WI) and Cameca (France) are now the sole commercial developers of APTs.

## Atomic Force Microscope

Atomic force microscopy is a technique for analysing the surface of a rigid material all the way down to the level of the atom. AFM uses a mechanical probe to magnify surface features up to 100 000 000 times, and produces 3D images of the surface. The technique is derived from a related technology, called scanning tunnelling microscopy (STM). The difference is that AFM does not require the sample to conduct electricity, whereas STM does. AFM also works in regular room temperatures, while STM requires special temperature and other conditions. AFM is being used to understand materials problems in many areas including data storage, telecommunications, biomedicine, chemistry, and aerospace.

The atomic force microscope was invented in 1986. It uses various forces that occur when two objects are brought within nanometres of each other. An AFM can work either when the probe is in contact with a surface, causing a repulsive force, or when it is a few nanometres away, where the force is attractive.

## Atomic Mass Unit (A.M.U.)

A unit of mass used to express relative atomic masses. It is equal to 1/12 of the mass of an atom of the isotope carbon-12 and is equal to $1.66033 \times 10^{-27}$ kg

The number of protons within the atomic nucleus of a chemical element.

### Atomistic Simultations

Atomic motion computer simulations of macromolecular systems are increasingly becoming an essential part of materials science and nanotechnology. Recent advances in supercomputer simulation techniques provide the necessary tools for performing computations on nanoscale objects containing as many as 300,000 atoms and on materials simulated with 1,000,000 atoms.

This new capability will allow computer simulation of mechanical devices or molecular machines using nanometer size components.

### ATPase

A protein complex responsible for converting electical potential energy (from a nanoscale batery) into ATP, the molecule used for immediate fuel by virtually all living systems.

### Attitude Control

Stabilizing a satellite's attitude (direction) in its orbit. Attitude control can be done by spinning the satellite, or by having it remain stabilized in three axes using an internal gyroscope and thrusters.

### Attitude

The position in space of a spacecraft or aircraft. A satellite's attitude can be measured by the angle the satellite makes with the object it is orbiting, usually the Earth. Attitude determines the direction a satellite's instruments face. The attitude of a satellite must be constantly maintained; this is known as attitude control.

### Aurora Borealis

Also known as the northern lights. They are streams of coloured light that appear in the northern night sky, often in the winter. They are caused by disturbances in the ionosphere.

### Austenite

Face-centered cubic iron; also iron and steel alloys that have the FCC crystal structure.

### Autogenous Control

In medical nanorobotics, the conscious control of in vivo nanorobotic systems by the human user or patient; in biochemistry, the action of a gene product that either inhibits (negative autogenous control) or activates expression of the gene coding for it.

### Automated Chemistry

The use of an automated machine (e.g. computer driven robot or fluidic devices) to carry out chemical reactions, purification and molecular assembly.

### Automated Engineering

The use of computers to perform engineering design, ultimately generating detailed designs from broad specifications with little or no human help. Automated engineering is a specialized form of artificial intelligence.

### Automated Manufacturing

As used here, nanotechnology-based manufacturing requiring little human labor.

### Automation

Automating processes is often a critical part of industrializing processes developed in the research lab. Higher throughput, quality control and better reproducibility are part of this process.. Automation may be cheaper, particularly in the long run.

### Autoproductivity

The ability of a system, under external control, to automatically produce an identical copy of itself.

# B

### Ballistic Magnetoresistance

Ballistic Magnetoresistance is yet another way in which spin orientation, encoding information on a storage medium such as a hard drive, can modify electrical resistance in a nearby circuit, thereby accomplishing the sensing of that orientation.

### Band Gap Energy

For semiconductors and insulators, the energies that lie between the valence and conduction bands; for intrinsic materials, electrons are forbidden to have energies within this range.

### Bandwidth

Bandwidth is the difference between the upper and lower cutoff frequencies of, for example, a filter, a communication channel, or a signal spectrum, and is typically measured in hertz. In case of a baseband channel or signal, the bandwidth is equal to its upper cutoff frequency. Bandwidth in hertz is a central concept in many fields, including electronics, information theory, radio communications, signal processing, and spectroscopy.

In computer networking and other digital fields, the term *bandwidth* often refers to a data rate measured in bits per second, for example network throughput. The reason is that according to Hartley's law, the digital data rate limit (or channel capacity) of a physical communication link is related to its bandwidth in hertz, sometimes denoted *analog bandwidth*.

For *bandwidth* as a computing term, less ambiguous terms are bit rate, throughput, goodput or channel capacity.

Bandwidth is a key concept in many telephony applications. In radio communications, for example, bandwidth is the range of frequencies occupied by a modulated carrier wave, whereas in optics it is the width of an individual spectral line or the entire spectral range.

In many signal processing contexts, bandwidth is a valuable and limited resource. For example, an FM radio receiver's tuner spans a limited range of frequencies. A government agency (such as the Federal Communications Commission in the United States) may apportion the regionally available bandwidth to licensed broadcasters so that their signals do not mutually interfere. Each transmitter owns a slice of bandwidth, a valuable (if intangible) commodity. For different applications there are different precise definitions. For example, one definition of bandwidth could be the range of frequencies beyond which the frequency function is zero. This would correspond to the mathematical notion of the support of a function (i.e., the total "length" of values for which the function is nonzero). A less strict and more practically useful definition will refer to the frequencies where the frequency function is *small*. Small could mean less than 3 dB below (i.e., less than half of) the maximum value, or more rarely 10 dB below, or it could mean below a certain absolute value. As with any definition of the *width* of a function, many definitions are suitable for different purposes.

### Basal Body

Short cylindrical array of microtubules plus their associated proteins found at the base of a eucaryotic cell cilium or flagellum. Serves as a nucleation site for the growth of the axoneme. Closely similar in structure to a centriole.

### Basal Lamina (plural basal laminae)

Thin mat of extracellular matrix that separates epithelial sheets, and many types of cells such as muscle cells or fat cells, from connective tissue. Sometimes called a basement membrane.

### Basal

Situated near the base. The

basal surface of a cell is opposite the apical surface.

### Base Pair

Two nucleotides in an RNA or a DNA molecule that are paired by hydrogen bonds—for example, G with C and A with T or U.

### Base

Molecule (usually containing nitrogen) that accepts a proton in solution. Often used to refer to the purines and pyrimidines in DNA and RNA.

### Basophil

White blood cell that releases histamine in an inflammatory response. Closely related to a mast cell.

### Bbionanotechnology

Includes molecular motors, biomaterials, single molecule manipulation technologies, biochip technologies, etc.

### Bearing

A mechanical device that permits the motion of a component (ideally, with minimal resistance) in one or more degrees of freedom while resisting motion (ideally, with a stiff restoring force) in all other degrees of freedom.

### BECON Bioengineering Consortium

The focus of bioengineering activities at the NIH. The Consortium consists of senior- level representatives from all of the NIH institutes, centers, and divisions plus representatives of other Federal agencies concerned with biomedical research and development. BECON is administered by the National Institute of Biomedical Imaging and Bioengineering (NIBIB).

### Bilayer Lipid Membrane (BLM)

The structure found in most biological membranes, in which two layers of lipid molecules are so arranged that their hydrophobic parts interpenetrate, whereas their hydrophilic parts form the two surfaces of the bilayer.

### Binary

Numbering system based on powers of 2 using only the digits 0 and 1, called "bits".

### Binding Energy

The reduction in the free energy of a system that occurs when a ligand binds to a receptor. Generally used to describe the total energy required to remove something, or to take a system apart into its constituent particles—for example, to separate two atoms from one another, or to separate an atom into electrons and nuclei.

### Binding Site

The active region of a receptor; any site at which a chemical species of interest tends to bind.

### Binding

The process by which a molecule (or ligand) becomes bound, that is, confined in position (and often orientation) with respect to a receptor. Confinement occurs because structural features of the receptor create a potential well for the ligand; van der Waals and electrostatic interactions commonly contribute.

### Biocatalysis

The application, both actual and potential, of biological catalysts (including whole cells or isolated components thereof, natural and modified enzymes and catalytic antibodies) for the synthesis, interconversion or degradation of chemical species. In addition to papers describing the synthetic applications of biotransformations the journal particularly welcomes papers focusing on the mechanistic principles, kinetics and thermodynamics of biocatalytic processes, the chemical or genetic modification of biocatalysts and the activity and stability of biocatalysts in non- aqueous and multi- phasic environments, including the design of large scale biocatalytic processes. The scope of the journal also encompasses biomimetic systems and environmental applications of biocatalysis where the mechanistic principles are understood or novel biocatalytic activities are involved.

### Biochauvinism

The prejudice that biological systems have an intrinsic superiority that will always give them a monopoly on self-reproduction and intelligence.

### Biocompatible Coated Materials

Biocompatible materials usually used in dental and bone implants that enhance biologic fixation, thereby increasing the bond strength between the coated material and bone, and minimize possible biological effects that may result from the implant itself. MeSH, 1999

## Biocompatible Materials

Synthetic or natural materials, other than drugs, that are used to replace or repair any body tissue or bodily function. MeSH, 1973

## Biodefense

The need for improved national defenses against biological attacks are urgently needed. A real-time reporting system needs to be developed, whereby officials can

be informed about an emerging threat before an agent has a chance to affect thousands of people. The development of integrated systems for detecting and monitoring biological agents is instrumental to this goal. The development of vaccines and novel detection platforms will enable preparedness, while also benefiting human health.

## Bioengineering

Is rooted in physics, mathematics, chemistry, biology, and the life sciences. It is the application of a systematic, quantitative, and integrative way of thinking about and approaching the solutions of problems important to biology, medical research, clinical proactive, and population studies.

The NIH Bioengineering Consortium agreed on the following definition for bioengineering research on biology, medicine, behavior, or health recognizing that no definition could completely eliminate overlap with other research disciplines or preclude variations in interpretation by different individuals and organizations.

Integrates physical, chemical, or mathematical sciences and engineering principles for the study of biology, medicine, behavior, or health. It advances fundamental concepts, creates knowledge for the molecular to the organ systems levels, and develops innovative biologics, materials, processes, implants, devices, and informatics approaches for the prevention, diagnosis, and treatment of disease, for patient rehabilitation, and for improving health.

## Biofabrication

Includes nanoparticle delivery systems, biomaterials, tissue engineering, implants and prostheses. [BECON, Nanoscience and Nanotechnology: Using biological processes to synthesize and manufacture chemicals and materials of high value to the Department of Defense.

Biological processes are characterized by: low energy barriers (~10Kcal, high temperatures/pressures not required); high reaction-, regio- and stereo- specificity (protection/deprotection wastes and catalyst poisoning avoided); spatiotemporal control of materials synthesis of defined composition and size with angstrom-level precision; and local control of the dielectric environment (largely eliminating the need for toxic solvents). Potential target materials and chemicals include composites with enhanced mechanical properties, ultra-low k dielectrics and thermoelectrics, optoelectronic materials, photonic devices (waveguides and logic elements), electronic materials (e.g., GaN, InGaN, AlGaN), elastomers, energetic materials, and adhesives. Biofabrication, DARPA.

## Bioinformatics

Bioinformatics is a field which uses computers to store and analyze molecular biological information. Using this information in a digital format, bioinformatics can then solve problems of molecular biology, predict structures, and even simulate macromolecules. In a more general sense, bioinformatics may be used to describe any use of computers for the purposes of biology, but the molecular-biology specific definition is by far the most common. At the beginning of the 21st century, scientists began sequencing entire species' genomes and storing them on computers, allowing for the use of bioinformatics to model and track a number of fascinating things.

One of these applications is in deducing evolutionary change in a species. By examining a genome and watching how it changes over time, evolutionary biologists can actually trace evolution as it occurs. The most well-known application of bioinformatics is sequence analysis. In sequence analysis, DNA sequences of various organisms are stored in databases for easy retrieval and comparison. The well-reported Human Genome Project is an example of sequence analysis bioinformatics. Using massive computers and various methods of collecting sequences, the entire human genome was sequenced and stored within a structured database. DNA sequences used for bioinformatics can be collected in a number of ways. One method is to go through a genome and search out individual sequences to record and store. Another method is to simply grab huge amounts of fragments and compare them all, finding whole sequences by overlapping the redundant segments. The latter method, known as shotgun sequencing, is currently the most popular because of its ease and speed.

By comparing known sequences of a genome to specific mutations, much information can be gleaned about undesirable mutations such as cancers. With the completed mapping of the human genome, bioinformatics has become very important in the research of cancers in the hope of an eventual cure. Computers are also used to collect and store broader data about species. The Species 2000 project, for example, aims to collect a large amount of information about every species of plant, fungus, and animal on the earth.

This information can then be used for a number of applications, including tracking changes in populations and biomes. There are many other applications of bioinformatics, including predicting entire protein strands, learning how genes express themselves in various species, and building complex models of entire cells.

### Biomaterials

Synthetic or natural materials that can replace or augment tissues, organs or body functions.

### Biomedical Nanotechnology

The science and technology of diagnosing, treating, and preventing disease and traumatic injury, of relieving pain, and of preserving and improving human health, using molecular tools and molecular knowledge of the human body.

## BioMEMS

Miniaturization engineering or MEMS applied to biotechnology or medicine. In BioMEMS the number of materials involved is much larger than in a comparable electronics application. Both instruments and sensors are used in BioMEMS. Applications include: Forensic science (e.g. O.J.s DNA); Clinical diagnostics (e.g. glucose in blood); Product development (e.g. new drug); and Quality control (e.g. pH of swimming pools).

## Biomimetics

The design of systems, materials, and their functionality to mimic nature. Current examples include layering of materials to achieve the hardness of an abalone shell or understanding why spider silk is stronger than steel.

## Biomimetic Chemistry

Knowledge of biochemistry, analytical chemistry, polymer science, and biomimetic chemistry is linked and applied to research in designing new molecules, molecular assemblies, and macromolecules having biomimetic functions. These new bio-related materials of high performance, including, for example, enzyme models, synthetic cell membranes, and biodegradable polymers, are prepared, tested, and constantly improved in this division for industrial scale production.

## Biomimetic Materials

Materials fabricated by *biometics* techniques, i.e., based on natural processes found in biological systems.

## BIOML Biopolymer Markup Language

Designed by the BIOML core team at Proteometrics, LLC and Proteometrics Canada Ltd. It is to be used "for the annotation of biopolymer sequence information. BIOML allows the full specification of all experimental information known about molecular entities composed of biopolymers, for example, proteins and genes."

## Biomolecular Materials

An emerging discipline, materials whose properties are abstracted from biology. They share many of the characteristics of biological materials but are not necessarily of biological origin. For example, they may be inorganic materials that are organized or processed in a biomimetic fashion. A key feature of biological and biomolecular materials is their ability to undergo *self- assembly*.

## Biomolecular Nanotechnology

Nanotechnology is abundant in nature in the form of biological systems; thus, nanoscientists have a wide variety of components and techniques already available. Biomolecular nanotechnology seeks to use biomolecules, biomolecular self-assembly & biomineralization to develop nanotechnology.

## Biomotors

Driven by energy sources such as *adenosine triphosphate (ATP)* for chemical transduction and other processes. These biomotors are considered to be biomolecular and are discussed in the body of this report, but strictly speaking they do not conform to the panel's definition of self- assembly.

## BioNEMS

BioNEMS is the Biofunctionalized nanoelectromechanical systems.

## Biopolymers

Macromolecules (including proteins, nucleic acids and polysaccharides) formed by living organisms.

## Biorobotics

Our research focuses on the role of sensing and mechanical design in motor control, in both robots and humans. This work draws upon diverse disciplines, including biomechanics, systems analysis, and neurophysiology. The main approach is experimental, although analysis and simulation play important parts. In conjunction with industrial partners, we are developing applications of this research in biomedical instrumentation, teleoperated robots, and intelligent sensors.

## Biosensor

The term "biosensor" is a general designation that denotes

either a sensor to detect a biological substance or a sensor which incorporates the use of biological molecules such as antibodies or enzymes. Biosensors are a subcategory of chemical sensors.

### Biostasis

A condition in which an organism's cell and tissue structure are preserved, allowing later restoration by cell repair machines. Applicable to cryonics.

### Biotin

A vitamin of the B complex. It is a co-enzyme for various enzymes that catalyse the incorporation of carbon dioxide into various compounds. It is essential for the metabolism of fats. Biotin is attached to pyruvate carboxylase by a long, flexible chain like that of lipoamide in the pyruvate dehydrogenase complex. Adequate amounts are normally produced by the intestinal bacteria in animals.

### Biotinylated-DNA

A DNA molecule labelled with biotin by incorporation of biotinylated -dUTP into a DNA molecule. It is used as a non-radioactive probe in hybridization experiments, such as Southern transfer. The detection of the labelled DNA is achieved by complexing it with streptavidin (an antibiotic with a high affinity for biotin) to which is attached a colour- generating agent such as horseradish peroxidase that gives a fluorescent green colour upon reaction with various organic reagents. Biotinylation: To label a probe with biotin.

### Biotransformation

The process whereby a substance is changed from one chemical to another *(transformed)* by a chemical reaction within the body. Metabolism or metabolic transformations are terms frequently used for the biotransformation process. However, metabolism is sometimes not specific for the transformation process but may include other phases of toxicokinetics. Biotransformation is vital to survival in that it transforms absorbed nutrients *(food, oxygen, etc.)* into substances required for normal body functions. For some pharmaceuticals, it is a metabolite that is therapeutic and not the absorbed drug. ... Biotransformation also serves as an important defense mechanism in that toxic xenobiotics and body wastes are converted into less harmful substances and substances that can be excreted from the body.

### Biovorous

From "biovore;" an organism capable of converting biological material into energy for sustenance.

### Bipolar-junction transistor

Transistor with n-type and p-type semiconductors having base-emitter and collector-base junctions.

### Bit

Numbering system based on powers of 2 using only the digits 0 and 1, called "bits".

### Black Hole

A region of space where gravity is so strong that nothing, not even light, can escape. Black holes with over a million times the sun's mass are found in quasars.

### Blue Goo

Opposite of Grey goo. Benificial tech, or "police" nanobots.

### Body

The non-technical term for a satellite's bus.

### Body Centered Cubic (BCC)

A common crystal structure found in some elemental metals. Within the cubic unit cell, atoms are located at corner and cell center positions.

### Bogosity Filter

A mechanism for detecting bogus ideas and propositions.

### Bond

Two atoms are said to be bonded when the energy required to separate them is substantially larger than the van der Waals attraction energy. Ionic bonds result from the electrostatic attraction between ions; covalent and metallic bonds result from the sharing of electrons among atoms; hydrogen bonds are weaker and result from dipole interactions and limited electron sharing. When used without modification, "bond" usually refers to a covalent bond.

## Bone Substitutes

Synthetic or natural materials for the replacement of bones or bone tissue. They include hard tissue replacement polymers, natural coral, hydroxyapatite, beta- tricalcium phosphate, and various other biomaterials. The bone substitutes as inert materials can be incorporated into surrounding tissue or gradually replaced by original tissue.

## Born-Oppenheimer Approximation

Permits the use of classical mechanics in modeling and thinking about molecular and atomic motions. Needless to say, this greatly simplifies the conceptual framework required for thinking about molecular machines. Once used as an argument on why MNT could not work.

## Bose-Einstein Condensates

Bose-Einstein Condensates "...aren't like the solids, liquids and gases that we learned about in school. They are not vaporous, not hard, not fluid. Indeed, there are no ordinary words to describe them because they come from another world — the world of quantum mechanics."

## Bottom Up

Building larger objects from smaller building blocks. Nanotechnology seeks to use atoms and molecules as those building blocks. The advantage of bottom-up design is that the covalent bonds holding together a single molecule are far stronger than the weak. Mostly done by chemists, attempting to create structure by connecting molecules.

## Bottom-up Fabrication

Building a small structure by assembling it out of atoms or small molecules.

## Brazilsat

Brazil contracted the Canadian firm Spar Aerospace Ltd. to build Brazilsat 1 and 2. This series of satellites was intended to give Brazil the ability to communicate within the entire country by satellite for the first time. Brazilsat 1 was launched on February 8, 1985 on a French Ariane 3 rocket.

## Brazing

A metal joining technique that uses a molten filler metal alloy having a melting temperature greater than about 425°C (800°F).

## Breakdown

Failure of a material resulting from an electrical overload. The resulting damage may be in the form of thermal damage (melting or burning) or electrical damage (loss of polarization in piezoelectric materials). The failing of a physical principle or unit of measurement as the scale is changed.

## Broadband Mode

In this mode of operation the error signal E representing bending of the cantilever not corrected by the feedback is scaled and added to the Z signal to provide a mathematically correct image (I=E+Z).

## Brownian Assembly

Brownian motion in a fluid brings molecules together in various position and orientations. If molecules have suitable complementary surfaces, they can bind, assembling to form a specific structure. Brownian assembly is a less paradoxical name for self-assembly.

## Brownian Motion

Motion of a particle in a fluid owing to thermal agitation . Motion of a particle in a fluid owing to thermal agitation, observed in 1827 by Robert Brown. Stochastic motion of a particle suspended in a surrounding gas or liquid of other particles, molecules or atoms, which is in thermodynamical equilibrium.

The motion of the particle is due to the collisions with the particles from its surrounding and as such is 'random', i.e. the change of particle's speed in any of the encounters is random. Although the Brownian motion is stochastical, it has some statistical properties which are fixed.

## Buckminsterfullerene

A fullerene with the chemical formula ($C_{60}$). A broad term covering the variety of buckyballs and carbon nanotubes that exist. Named after the architect Buckminster Fuller, who is famous for the geodesic dome, which buckyballs resemble.

## Buckyball

The buckyball (fullerenes) are a recently-discovered family of carbon allotropes named after Buckminster Fuller. They are molecules composed entirely of carbon, in the form of a hollow sphere, ellipsoid, or tube. Spherical fullerenes are sometimes called buckyballs, the $C_{60}$ variant is often compared to the typical white and black soccer football, the Telstar (football) of 1970. Cylindrical fullerenes are called buckytubes. Fullerenes are similar in structure to graphite, which is composed of a sheet of linked hexagonal rings, but they contain pentagonal (or sometimes heptagonal) rings that prevent the sheet from being planar. In molecular beam experiments, discrete peaks were observed corresponding to molecules with the exact mass of sixty or seventy or more carbon atoms. In 1985, Harold Kroto (of the University of Sussex), James Heath, Sean O'Brien, Robert Curl and Richard Smalley, from Rice University, discovered $C_{60}$, and shortly after came to discover the fullerenes. Kroto, Curl, and Smalley were awarded the 1996 Nobel Prize in Chemistry for their roles in the discovery of this class of compounds. $C_{60}$ and other fullerenes were later noticed occurring outside of a laboratory environment (e.g. in normal candle soot). By 1991, it was relatively easy to produce grams of fullerene powder using the techniques of Donald Huffman and Wolfgang Krätschmer.

Fullerene purification remains a challenge to chemists and determines fullerene prices to a large extent. So called endohedral fullerenes have ions or small molecules incorporated inside the cage atoms. Fullerene is an unusual reactant in many organic reactions such as the Bingel reaction discovered in 1993.

Buckminsterfullerene ($C_{60}$) was named after Richard Buckminster Fuller, a noted architect who popularized the geodesic dome. Since buckminsterfullerenes have a similar shape to that sort of dome, the name was thought to be appropriate. As the discovery of the fullerene family came *after* buckminsterfullerene, the name was shortened to illustrate that the latter is a type of the former.

Buckminsterfullerene (IUPAC name ($C_{60}$-$I_h$) fullerene) is the smallest fullerene in which no two pentagons share an edge. It is also the most common in terms of natural occurrence, as it can often be found in soot.

The structure of $C_{60}$ is a truncated icosahedron, which resembles a round soccer ball of the type made of hexagons and pentagons, with a carbon atom at the corners of each hexagon and a bond along each edge. The $C_{60}$ molecule has two bond lengths. The 6:6 ring bonds (between two hexagons) can be considered "double bonds" and are shorter than the 6:5 bonds (between a hexagon and a pentagon).

For the past decade, the chemical and physical properties of fullerenes have been a hot topic in the field of research and development, and are likely to continue to be for a long time. In April 2003, fullerenes were under study for potential medicinal use: binding specific antibiotics to the structure to target resistant bacteria and even target certain cancer cells such as melanoma.

The October 2005 issue of Chemistry and Biology contains an article describing the use of fullerenes as light-activated antimicrobial agents.

In the field of nano-technology, heat resistance and superconductivity are some of the more heavily studied properties.

A common method used to produce fullerenes is to send a large current between two nearby graphite electrodes in an inert atmosphere. The resulting carbon plasma arc between the electrodes cools into sooty residue from which many fullerenes can be isolated.

In chemistry, fullerenes are stable, but not totally unreactive. The $sp^2$-hybridized carbon atoms, which are at their energy minimum in planar graphite, must be bent to form the closed sphere or tube, which produces angle strain. The characteristic reaction of fullerenes is electrophilic addition at 6,6-double bonds, which reduces angle strain by changing $sp^2$-hybridized carbons into $sp^3$-hybridized ones.

The change in hybridized orbitals causes the bond angles to decrease from about 120 degrees in the $sp^2$ orbitals to about 109.5 degrees in the $sp^3$ orbitals. This decrease in bond angles allows for the bonds to bend less when closing the sphere or tube, and thus, the molecule becomes more stable.

Other atoms can be trapped inside fullerenes to form inclusion compounds known as endohedral fullerenes. Recent evidence for a meteor impact at the end of the Permian period was found by analysing noble gases so preserved. Metallofullerene-based inoculates using the rhonditic steel process are beginning production as one of the first commercially-viable uses of buckyballs.

# C

## Cadherin

Cadherins are a class of proteins which are expressed on the surface of cells. They play important roles in cell adhesion whereby they ensure cells within tissues are bound together. They are dependent on calcium ($Ca^{2+}$) ions to function, hence their name. The cadherin superfamily includes cadherins, protocadherins, desmogleins and desmocollins, and more. Structurally, they share "cadherin repeats", which are the extracellular $Ca^{2+}$ binding domains.

There are multiple classes of cadherin molecule, each designated with a one-letter prefix (generally noting the type of tissue with which it is associated). Cadherins within one class will only bind to themselves. For example, an N-cadherin will bind only to another N-cadherin molecule. Because of this specificity, groups of cells that express the same type of cadherin molecule tend to cluster together during development, while cells expressing different types of cadherin molecules tend to separate.

E-cadherin (epithelial) is probably the best understood cadherin. It consists of 5 cadherin repeats (EC1 ~ EC5) in the extracellular domain, one transmembrane domain, and an intracellular domain that binds p120-catenin and beta-catenin. The intracellular domain contains a highly phosphorylated region vital to beta-catenin binding and therefore to E-cadherin function. Beta-catenin can also bind to alpha-catenin. Alpha-catenin participates in regulation of actin-containing cytoskeletal filaments. In epithelial cells, E-cadherin-containing cell-to-cell junctions are often adjacent to actin-containg filaments of the cytoskeleton.

E-cadherin is first expressed in the 2-cell stage of mammalian development, and becomes phosphorylated by the 8-cell stage, where it causes compaction. In adult tissues, E-cadherin is expressed in epithelial tissues, where it is constantly regenerated with a 5 hour half-life on the cell surface.

Loss of E-cadherin function or expression has been implicated in cancer progression and metastasis. E-cadherin downregulation decreases the strength of cellular adhesion within a tissue, resulting in an increase in cellular motility. This in turn may allow cancer cells to cross the basement membrane and invade surrounding tissues.

## Calibration

Calibration refers to the process of setting the magnitude of the output (or response) of a measuring instrument to the magnitude of the input property or attribute within specified accuracy and precision. For example, a thermometer could be calibrated so that it showed the temperature in Celsius at the correct point.

For physical constants, weights, and measures, there are known and agreed values in the International System of Units (SI). Such constants include the length of the metre, the mass of the kilogram, and the volume of a litre.

In the United States, the National Institute of Standards and Technology, a part of the federal government, maintains standards and is considered the arbiter and ultimate (in the U.S.) authority for values of SI units and industrial standards. NIST also defines traceability, by which an instrument's accuracy is established in an unbroken chain relating an instrument's measurements through one or more derivative standards to a standard maintained by NIST, as well as uncertainty.

In science, a calibrated test tube is one with measurements up the side. In computing, an interactive whiteboard pen or other input method can be calibrated so that it moves the cursor to the correct point on the screen.

## Calvin Cycle

The Calvin cycle (or Calvin-Benson cycle or carbon fixation) is a series of biochemical reactions that takes place in the stroma of chloroplasts in photosynthetic organisms. It was discovered by Melvin Calvin and Andrew Benson at the University of California, Berkeley. James Bassham also made important contributions to elucidating this pathway. It is one of the light-independent reactions.

During photosynthesis, light energy is used to generate chemical free energy, stored in ATP and NADPH. The light-independent Calvin cycle, also (misleadingly) known as the "dark reaction" or "dark stage", uses the energy from short-lived electronically-excited carriers to convert carbon dioxide and water into organic compounds that can be used by the organism (and by animals which feed on it). This set of reactions is also called *carbon fixation*. The key enzyme of the cycle is called RuBisCO. In the following equations, the chemical species (phosphates and carboxylic acids) exist in equilibria among their various ionized states as governed by the pH. The enzymes in the Calvin cycle are functionally equivalent to many enzymes used in other metabolic pathways such as glycolysis and gluconeogenesis, but they are to be found in the chlorophlast stroma instead of the cell cytoplasm, separating the reactions. They are activated in the light (which is why the name "dark reaction" is misleading), and also by products of the light-dependent reaction.

These regulatory functions prevent the Calvin cycle from operating in reverse to respiration, which would create a continuous cycle of carbon dioxide being reduced to carbohydrates, and carbohydrates being respired to carbon dioxide. Energy (in the form of ATP) would be wasted in carrying out these reactions that have no net productivity. The two G3P molecules (or one F6P molecule) which have exited the cycle are used to make carbohydrates. In simplified versions of the Calvin cycle they may be converted to F6P after exit, but this conversion is also part of the cycle. Hexose isomerase converts about half of the F6P molecules into glucose-6-phosphate. These are dephosphorylated and the glucose can be used to form starch, which is stored in, for example, potatoes, or cellulose used to build up cell walls. Other glucose, with fructose, forms sucrose, the plant sugar.

**Cam**

A component that translates or rotates to move a contoured surface past a follower; the contours impose a sequence of motions (potentially complex) on the follower.

**cAMP**

Enzyme that phosphorylates target proteins in response to a rise in intracellular cyclic AMP. First identified in skeletal muscle as part of the pathway of regulation of glycogen breakdown in response to adrenaline.

**Canadarm**

The nickname for Spar Aerospace's SRMS (Space shuttle Remote Manipulator System). The first Canadarm was a gift from Canada to NASA's space shuttle program; NASA subsequently bought Canadarms to equip the rest of the shuttle fleet. The Canadarm is used during shuttle flights to release satellites into orbit, retrieve them if they malfunction, and aid in their repair.

**Cantilever Vibration Mode**

An atomic force microscope (AFM) combined with ultrasonic frequency vibration of a cantilever excited at its support. This method enables both topography and elasticity imaging of stiff samples such as metals and ceramics, without a need for bonding a transducer to the sample.

**Capacitance**

The charge-storing ability of a capacitor, defined as the magnitude of charge stored on either plate divided by the applied voltage. A 1-F capacitor charged to 1 V contains C of charge and 1 C is an amount of charge equal to that of about $6.24 \times 10^{18}$ electrons.

**Capillaries**

Microscopic blood vessels that carry oxygenated blond to tissues.

**Carbanion**

A highly reactive anionic

chemical species with an even number of electrons and an unshared pair of electrons on a tetravalent carbon atom.

### Carbene

A highly reactive chemical species containing an electrically neutral, divalent carbon atom with two nonbonding valence electrons; a prototype is $CH_2$.

### Carbon Black

Carbon black is a powdered form of elemental carbon. The primary use of carbon black is in rubber products, mainly tyres and other automotive products, but also in many other rubber products such as hoses, gaskets and coated fabrics. Much smaller amounts of carbon black are used in inks and paints, plastics and in the manufacture of dry-cell batteries.

### Carbon Cycle

The carbon cycle is the means by which carbon atoms are exchanged between living things, the ground, the oceans, and the skies; or biosphere, geosphere, hydrosphere, and atmosphere, respectively. There is about 1,000,000 gigatons of carbon on Earth, most of which is locked in sedimentary rocks and never reaches the surface. At the surface, carbon is continuously engaging in a dynamic exchange of consumption and production. This active exchange is referred to as the carbon cycle. The biosphere contains approximately 600 gigatons of carbon, representing the most condensed carbon pool on Earth. Through the carbon cycle, carbon circulates in and out of the biosphere as plants and animals respirate, excrete, perish, and perform photosynthesis.

Soil in the form of organic matter, which plants and animals become when they die, contains about 1,500 gigatons of carbon. Some of this carbon sinks deep into the Earth into sedimentary rocks, never to surface again. Some of it reaches the oceans and dissolves. The deep oceans contain a substantial amount of carbon, about 38,000 gigatons About 5.5 gigatons of fossil fuel emissions are released into the atmosphere each year by human industry, creating a figurative top-heavy carbon cycle.

International coalitions have been formed to limit these emissions, with moderate success so far. Remaining coal deposits total about 3,000 gigatons of carbon, and remaining oil and gas reserves make up about 300

## Carbon Nanofoam

Carbon nanofoam is the fifth known allotrope of carbon discovered in 1997 by Andrei V. Rode and co-workers at the Australian National University in Canberra. It consists of a low-density cluster-assembly of carbon atoms strung together in a loose three-dimensional web. Each cluster is about 6 nanometers wide and consists of about 4000 carbon atoms linked in graphite-like sheets that are given negative curvature by the inclusion of heptagons among the regular hexagonal pattern.

This is the opposite of what happens in the case of buckminsterfullerenes, in which carbon sheets are given positive curvature by the inclusion of pentagons. The large-scale structure of carbon nanofoam is similar to that of an aerogel, but with 1% of the density of previously produced carbon aerogels — only a few times the density of air at sea level. Unlike carbon aerogels, carbon nanofoam is a poor electrical conductor. The nanofoam contains numerous unpaired electrons, which Rode and colleagues propose is due to carbon atoms with only three bonds that are found at topological and bonding defects. This gives rise to what is perhaps carbon nanofoam's most unusual feature; it is attracted to magnets, and below "183 °C can itself be made magnetic. This property of ferromagnetism has also been seen in other allotropes of carbon including fullerene subjected to high pressures and temperatures and graphite irradiated with high energy protons.

A spongy solid that is extremely lightweight and, unusually, attracted to magnets... John Giapintzakis of the University of Crete has used an electron microscope to study the structure of the nanofoam. He says it is the fifth form of carbon known after graphite, diamond and two recently discovered types: hollow spheres, known as buckminsterfullerenes or buckyballs, and nanotubes. Electrically conductive carbon nanofoams are a new material with many of the properties of traditional aerogel material. These materials are available in the form of monoliths, granules, powders and papers.

They are synthetic, lightweight foams in which the solid matrix and pore spaces have nanometer- scale dimensions. Prepared by sol- gel methods, nanofoams typically have low density, continuous porosity, high surface area, and fine cell/ pore sizes. Carbon nanotubes are cylindrical structures made only from carbon atoms that are about 1 nm in diameter and 1-100 microns in length. In a graphite sheet, carbon atoms arrange themselves in a two-dimensional hexagonal lattice. Carbon nanotubes can be though of as a strip of graphite sheet that is rolled up to form a cylinder. There are many different ways to cut up a piece of graphite and roll it up to form a tube. The tubes can have different diameters and different chiralities. The chirality is the twist of the rows of atoms along the length of the tube. Sometimes the atom rows are parallel to the axis of the tube and sometimes the rows form a helix that winds along the tube. When one or more tubes grow inside another carbon nanotube, it is called a multiwalled nanotube. Carbon naotubes are very strong; they are 5 times as strong as steel for the same wieght.

The electrical properties of carbon nanotubes depend on their diameter and their chirality. Some tubes are metallic and some are semiconductors. Research on carbon nanotubes at TU Delft, Nanotech Now's nanotube and buckyball page. A form of carbon related to fullerenes, except that the carbon atoms form extended hollow tubes instead of closed, hollow spheres. Carbon nanotubes can also form as a series of nested, concentric tubes. Carbon nanotubes can be used as nanometer-scale syringe needles for injecting molecules into cells and as nanoscale probes for making fine-scale measurements. Carbon nanotubes can be filled and capped, forming nanoscale test tubes or potential drug delivery devices. Carbon nanotubes can also be "doped," or modified with small amounts of other elements, giving them electrical properties that include fully insulating, semiconducting, and fully conducting.

## Carbon Nanotubes

These tiny straw-like cylinders of pure carbon have useful electrical properties. They have already been used to make tiny transistors and one-dimensional copper wire. They were developed by using nanotechnology, a relatively new field that involves building electronic circuits and devices from single atoms and molecules. Nano means one thousand millionth of a unit. A nanometer is therefore one thousand millionth of a meter. The first nanofabrication experiments occurred in 1990 when individual xenon atoms were placed on a nickel substrate and used to spell out a company logo. One primary goal of nanotechnology is to build computer chips and other devices that are thousands of times smaller than they are now Carbon nanotubes have enormous theoretical possibilities but have not lived up to the hype surrounding their development. Researchers have continued to look for ways to use them, however, as successful applications have the potential to be highly lucrative.

Scientists have recently succeeded in altering carbon nanotubes so that they supply electrons when exposed to light. This was done by having two flat rings of carbon molecules sandwich a ferrocene (iron) molecule. Ferrocene is known for its tendency to relinquish electrons. When exposed to visible light, the carbon atoms accepted the ferrocene molecule. This is the first time that carbon nanotubes have been hybridized to undergo light-induced electron transfer. Researchers say that these modified carbon nanotubes are the first step in building solar cells using this technology. The newly-discovered ability of carbon nanotubes to serve as electron sources has great potential. Carbon nanotubes may one day replace the metal filaments in X-ray machines, which tend to burn out quickly. Scientists hope to use them to develop portable X-ray machines for use in airport security, ambulances, and customs work.

Carbon nanotubes also have great significance for use in flat-panel displays, microwave generators, devices for electric *surge protection,* and high intensity lamps. Carbon nanotubes consist of tiny cylinders of graphite, closed at each end with caps which contain precisely six pentagonal rings. Since their discovery in 1991, physicists, chemists, and materials scientists have been interested in nanotubes because of their electronic properties, potential as nanotest-tubes, and incredible stiffness. Attempting to cover all the important areas of nanotube research, Harris (chemistry, Birmingham U., UK) considers the various methods for synthesizing nanotubes, including catalytically produced and single-walled tubes, and summarizes current thinking on growth mechanisms. Theoretical approaches to observation, methods of opening and filling nanotubes, electronic properties, and nanotube structures are discussed. Later chapters cover inorganic analogues of fullerenes and nanotubes, look at spheroidal forms of carbon, and consider future directions in which nanotube science might develop.

## Carbonium ion

A highly reactive cationic chemical species with an even number of electrons and an unoccupied orbital on a carbon atom.

## Carbonyl

A chemical moiety consisting of O with a double bond to C. If the C is bonded to N, the resulting structure is termed an amide; if it is bonded to O, it is termed a carboxylic acid or an ester linkage.

## Carboxylic Acid

A molecule that includes a C having a double bond to O and a single bond to OH.

## Catalysis

Chemical reaction between two substances aided by the presence of the third substance called *catalyst*. The catalyst material is unchanged by the reaction, but helps in creating intermediate complexes which promote the reaction. Catalysts can speed up chemical reactions or even enable reactions which were impossible without the presence of a catalyst material. It is known that catalysts specifically shaped on a scale of nanometer (nanostructured catalysts e.g. stepped surfaces of materials, zeolites, nanocrystals...) can additionally raise the efficiency of the reaction. This also has to do something with a very high active surface area of such materials. A prototypical example of catalysis is the oxidation of CO gas in the automobile exhaust fumes into $CO_2$, i.e. CO and $O_2$ molecules in the presence of catalyst material (nickel, tin, platinum, rhodium...) combine into $CO_2$ molecules. Recently, the interest in catalysis has been increased due to the necessity to produce hydrogen for some future 'clean, hydrogen based economy'. It is investigated how to produce hydrogen gas from e.g. biomass or methane by efficiently using catalyst materials. Catalysis is extremely important for the present day technology. Production of most chemicals, including fertilisers, relies on it.

## Catalyst

A substance that increases the rate of a chemical reaction by reducing the activation energy, but which is left unchanged by the reaction. A catalyst works by providing a convenient surface for the reaction to occur. The reacting particles gather on the catalyst surface and either collide more frequently with each other or more of the collisions result in a reaction between particles because the catalyst can lower the activation energy for the reaction.

## Cathode

The electrode in an electrochemical cell or galvanic couple at which a reduction reaction occurs; in other words the electrode receiving electrons from an external circuit.

## Cation

An ion consists of one or more atoms and carries a unit charge of electricity. Those that are positively electrified (hydrogen and the metals) are called cations (cf. anion).

## cDNA

Abbreviation for complementary deoxyribonucleic acid. cDNA has been copied directly from mRNA, the reverse of transcription. It differs from genomic DNA in the nucleus, since it does not contain the introns, or sequences upstream or downstream of the corresponding genes. cDNA libraries consist of samples from individual tissues, and will only contain the sequences transcribed in that tissue.

### Cell Engineering

Deliberate artificial modifications to biological cellular systems on a cell-by-cell basis.

### Cell Surgery

Nanorobotics, modifying cellular structures using medical nanomachines. Modifying cellular structures using medical nanomachines.

A pioneering form of surgery has been developed that can restore the sight of patients by using stem cells to encourage damaged eyes to repair themselves. A team of British specialists has successfully treated more than a dozen patients with impaired corneas by transplanting human stem cells grown in a laboratory on to their eyes. Recent operations on ten patients showed that the technique restored sight in seven cases of people who had been blinded after getting acid, alkali and boiling metal in their eyes, or because of congenital disorders. Many of the patients treated at the Centre for Sight, Queen Victoria Hospital, in East Grinstead, West Sussex, had been told that they had no hope of getting their sight back, or had already undergone failed corneal transplants. The process involves taking stem cells, which occur naturally in the eye, and developing them into sheets of cells in the laboratory.

These are transplanted on to the surface of the eye where they are held in place by an amniotic membrane, which dissolves away as the sheet fuses to the eye.

### Ceramic

A nonmetallic material made from clay and hardened by firing at high temperature; it contains minute silicate crystals suspended in a glassy cement.

### Cerium Oxide

There are numerous commercial applications for cerium including metallurgy, glass and glass polishing, ceramics, phosphors and catalysts. In catalysis, cerium is used in the form of cerium (IV) oxide, $CeO_2$. Cerium oxide is a highly stable, nontoxic, refractory ceramic material with a melting point of 2600°C and a density of 7.13 g.cm$^{-3}$. The crystal structure is fluorite face centred cubic with a lattice constant of 5.11Å. The rare earth oxides are generally some of the most thermally stable of all known materials and as such may be used in extremely high temperature applications without decomposition of the oxide.

### Cermet

Cermet is an abbreviation for "'ceramic" and "metal." CerMet is a composite material composed of ceramic (cer) and metallic (met) materials. A Cermet is ideally designed to have the optimal properties of both a ceramic, such as high temperature resistance and hardness, and those of a metal, such as the ability to undergo plastic deformation. The metal is used as a binder for an oxide, boride, carbide, or alumina. Generally, the metallic elements used are nickel, molybdenum, and cobalt. Depending on the physical structure of the material, cermets can also be metal matrix composites, but cermets are usually less than 20% metal by volume.

It is used in the manufacture of resistors (especially potentiometers), capacitors, and other electronic components which may experience high temperatures.

Some types of cermet are also being considered for use as spacecraft shielding as they resist the high velocity impacts of micrometeoroids and orbital debris much more effectively than more traditional spacecraft materials such as aluminum and other metals.

One application of these materials is their use in vacuum tube coatings which are key to solar hot water systems. Cermets are also used in dentistry as a material for fillings and prostheses.

### Charge Coupled Device

A highly sensitive camera, this counts individual photons of light to put together an image. It can take very detailed photographs of small areas.

### Charged Particle

A charged particle is fundamental particle (such as an electron) or a compound particle (such as an alpha particle) that carries a charge. Some common charged particles

### Chemistry

The science (and corresponding technology) that seeks to understand (and control) the atomic and molecular structure of matter.

### Chemotactic Nanosensor

In medical nanorobotics, a nanosensor used to determine the chemical characteristics of surfaces, possibly configured as a pad coated with an array of reversible, perhaps reconfigurable, artificial molecular receptors.

### Chip

A die (unpackaged semiconductor device) cut from a silicon wafer, incorporating semiconductor circuit elements such as a sensor, actuator, resistor, diode, tr nsistor, and/or capacitor.

### Chromatography

The physical method of separation in which the components to be separated are distributed between two phases, one of which is stationary while the other moves in a definite direction. Chromatography is a widely used for the separation, identification, and determination of the chemical components in complex mixtures.

### Chronocyte

In medical nanorobotics, a theorized mobile, mass-storage (nanorobotic) device, similar to a communicyte, that may be used as a mobile source of precisely synchronized universal time inside the human body.

### C-kinase

Ca2+-dependent protein kinase that, when activated by diacylglycerol and an increase in the concentration of Ca2+, phosphorylates target proteins on specific serine and threonine residues.

### Classical Mechanics

Classical mechanics describes a mechanical system as a set of particles (which in a limiting case can form continuous media) having a well-defined geometry at any given time, and undergoing motions determined by applied forces and by the initial positions and velocities of the particles. The forces themselves may have electromagnetic or quantum mechanical origins. Classical statistical mechanics uses the same physical model, but treats the geometry and velocities as uncertain, statistical quantities subject to random thermally-induced fluctuations. Classical mechanics and classical statistical mechanics give a good account of many mechanical properties and behaviors of molecules; but for describing the electronic properties and behaviors of molecules, they are often useless.

### Clays

A family of finely crystalline materials, composed primarily of silicates, with layered crystal structures. Clays may be refined from geological materials or may be synthetic.

### CMOS

An acronym for *complementary metal-oxide-semiconductor*, as in CMOS transistor and CMOS logic. Complementary metal oxide semiconductor—integrated circuit containing n-channel and p-channel MOSFETs.

### Cobots

Collaborative robots designed to work alongside human operators. Prototype cobots are being used on automobile assembly lines to help guide heavy components like seats and dashboards into cars so they don't damage auto body parts as workers install them.

### Cognotechnology

Convergence of nanotech, biotech and IT, for remote brain sensing and mind control.

### Coherence

From quantum physics, coherence is a property exhibited by matter and energy when it remains unmeasured and is thus able to exist in more than one state simultaneously (superposition).

### Col

In describing landforms, a pass between two valleys is sometimes termed a col. In describing molecular potential energy functions, this term is commonly used to describe analogous features of the PES; a col is the region around a saddle point having negative curvature along one axis and positive curvature along all orthogonal axes.

## Cold Plasma Analyzer

An instrument on board the Swedish scientific satellite Freja. By sweeping its sensor through plasma, it obtains measurements of the charged particles in a layer of the atmosphere, the magnetosphere and upper ionosphere, and thus helps scientists understand Earth's atmosphere and how to protect it from pollutants.

## Color Contacts

Color contact lenses are used to enrich or change the natural color of the eyes for movies, television, or for sheer fun. They can also be corrective lenses, but a corrective prescription is not necessary to wear color contacts. The iris is the portion of the eye that contains pigment. Color contacts are clear in the center to reveal the pupil, while the outer ring that rides on the iris is tinted, or in some lenses, hand painted. Hand painted lenses include subtle detail and varied colors that recreate depth in the iris, giving the eyes a natural appearance. Although very dark eyes appear to be one solid color, anyone with lighter brown, hazel, green or blue eyes actually have various colors and patterns in the iris.

Some have a starburst pattern; others have tiny rays of yellow, gold or black within the blue, green or gray background. These subtle patterns and varied colors are what give eyes a feeling of depth. The real difference in quality between color contacts is in how natural they look.

The more natural looking, the more expensive they will be compared to solid-tinted lenses. Making single-color contacts is a less expensive process than making intricate patterns of varied colors. Less expensive lenses, although providing striking color, can look unnatural. Prices therefore vary widely, depending on the lenses chosen. Color contacts will require an eye exam even if the lenses will not be corrective. An optometrist will need to make sure the lenses fit properly and center themselves after blinking.

A lens that fits too tight will dry the eye out and become uncomfortable, while a lens that is not tight enough will wander off the iris.

## Colorimetry

The methods used to measure color and to define the results of the measurements.

## Combination Products

Drug- device, drug- biologic, device- biologic, drug- device biologic.

## Combinatorial Biology

Involves genetic manipulation of bacteria and fungi that produce complex natural products. This technology includes construction of large libraries of recombinant microbes capable of generating novel organic molecules and engineering secondary metabolite biosynthetic pathways to modify valuable biologically active microbial metabolites.

## Combinatorial Library

Combinatorial Libraries & synthesis glossary Narrower terms: combinatorial peptide libraries, combinatorial protein libraries, compound library, hit optimization library, lead discovery library, biased libraries, combinatorial antibody libraries, directed libraries, focused libraries, pool, pool/ split libraries, sub- library, random libraries, unbiased librarie.

## Combinatorial Materials Design

Uses compμting power (sometimes together with massive parallel experimentation) to screen many different materials possibilities to optimize properties for specific applications.

## Combined Dynamic x Mode (CODYMode)

In CODYMode SFM at least two oscillations with sufficiently different frequencies and amplitudes are superimposed and interact with the sample surface. This enables the concurrent measurement of the topography, adhesive and frictional forces beside further mechanical surface properties of the sample.

## Communications Satellite

A type of satellite used for communications on Earth by allowing radio, television, and telephone transmissions to be sent live anywhere in the world. Before satellites, transmissions were difficult or impossible at long distances. The signals, which travel in straight lines, could not bend around the round Earth to reach a destination far away. Because communications satellites are in orbit, the signals can be sent instantaneously into space and then redirected to another satellite or directly to their destination.

## Communicyte

In medical nanorobotics, a theorized mobile, mass-storage (nanorobotic) device that can be used for information transport throughout the human body.

## Compliance

The reciprocal of stiffness; in a linear elastic system, displacement equals force times compliance.

## Complementary Metal-Oxide Semiconductor (CMOS)

The semiconductor technology used in the transistors that are manufactured into most of today's computer microchips.

## Composites

Combinations of metals, ceramics, polymers, and biological materials that allow multifunctional behaviour. One common practice is reinforcing polymers or ceramics with ceramic fibres to increase strength while retaining light weight and avoiding the brittleness of the monolithic ceramic.

Materials used in the body often combine biological and structural functions (e.g., the encapsulation of drugs).

## Computational Nanotechnology

Permits the modeling and simulation of complex nanometer-scale structures.

The predictive and analytical power of computation is critical to success in nanotechnology: nature required several hundred million years to evolve a functional "wet" nanotechnology; the insight provided by computation should allow us to reduce the development time of a working "dry" nanotechnology to a few decades, and it will have a major impact on the "wet" side as well.

### Computronium

A highly (or optimally) efficient matrix for computation, such as dense lattices of nanocomputers or quantum dot cellular automata.

### Conducting AFM

Contact mode of AFM operation with conductive tip. The tip-surface current is measured during scanning.

### Conductive

Allows electricity to travel through it well. *Conductors* are conductive.

### Conductor

A substance, material or object that allows electricity to flow through it with low resistance. Màterial such as the metals copper or aluminum that conducts electricity via the motion of electrons.

### Configuration Space

A mathematical space describing the three-dimensional configuration of a system of particles (e.g., atoms in a nanomechanical structure) as a single point; the configuration space for an $N$ particle system has $3N$ dimensions.

### Confocal Microscopy

A light microscopic technique in which only a small spot is illuminated and observed at a time. An image is constructed through point- by- point scanning of the field in this manner. Light sources may be conventional or laser, and fluorescence or transmitted observations are possible.

### Conformation

The shape of a molecule. A molecular geometry that differs from other geometries chiefly by rotation about single or triple bonds; distinct conformations (termed con formers) are associated with distinct potential wells. Typical biomolecules and products of organic synthesis can interconvert among many conformations; typical diamondoid structures are locked into a single potential well, and thus lack conformational flexibility.

### Conjugated

A conjugated pi system is one in which pi bonds alternate with single bonds. The resulting electron distribution gives the intervening single bonds partial double-bond character, the pi electrons become delocalized, and the energy of the system is reduced.

### Conjugation

In medical nanorobotics, the docking of two or more nanorobots for the purpose of exchanging information, energy or materials, or to establish a larger multirobotic structure; in biology, the union of two unicellular organisms accompanied by an interchange of nuclear material, as in Paramecium.

### Conservative

In design and analysis, a conservative model or a conservative assumption is one that departs from accuracy in such a way that it reduces the chances of a false-positive assessment of the feasibility of the system in question. Conservative assumptions overestimate problems and underestimate capabilities.

### Constant-amplitude Mode of Non-contact AFM

In constant-amplitude mode the tip is oscillating with a constant amplitude A of typically 1–10 nm at the eigenfrequency f of the cantilever, which may shift by Df due to forces between tip and sample. The oscillation amplitude is kept constant by a regulation circuit that excites a piezoactuator with a sinusoidal voltage of the oscillation frequency f and an amplitude Vexc . The actuator shakes the fixed end of the cantilever. When the cantilever oscillation is damped due to the tip-sample interaction, Vexc will increase to maintain the oscillation amplitude constant. By recording Df and Vexc simultaneously, forces and dissipation can be measured.

### Contact AFM (C-AFM)

In contact AFM modes of operation scanning probe tip touches with the investigated sample surface. During scanning the probe or sample can be vibrated with frequencies below resonance, in resonance (including overtones) and above resonance. Contact AFM includes several d.c.and a.c. modes: constant-height mode, constant-force mode, force modulation mode, scanning microdeformation microscopy, resonance contact scanning force microscopy, ultrasonic force microscopy (sample-induced and waveguide), heterodyne force microscopy, and so on.

### Contelligence

(Consciousness + intelligence) The combination of awareness and computational power required in an Artificially Intelligent network before we could, without loss of anything essential, upload ourselves into them.

### Convergent Assembly

A process of fastening small parts to obtain larger parts, then fastening those to make still larger parts, and so on; convergent assembly can be used to build a product from many, much smaller, components. "...rapidly make products whose size is measured in meters starting from building blocks whose size is measured in nanometers. It is based on the idea that smaller parts can be assembled into larger parts, larger parts can be assembled into still larger parts, and so forth.

### Copolymer

A polymer that consists of two or more dissimilar monomer units in combination in its molecular chains. Also a polymer formed from the polymerization of more than one type of monomer.

### Corrosion

Deteriorative loss of a metal as a result of dissolution environmental reactions.

### Cospas-Sarsat Satellites

Search and rescue satellites in polar orbit. The Cospas-Sarsat system is a partnership between several countries including the United States, Canada, and Russia. The first satellite in the Cospas-Sarsat system was launched in 1982, and by 1984 the system was operational.

### Coulomb [C]

Measure of electrical charge: 1 C is an amount of charge equal to that of about $6.24 \times 10^{18}$ electrons.

### Coupling

A connection between more than one object or energy pathway so that together they function as a single unit.

### Covalent Bond

A bond formed by sharing a

pair of electrons between two atoms. A primary interatomic bond that is formed by the sharing of electrons between neighboring atoms. An attachment between two atoms resulting from the sharing of a pair of electrons

### Covalent Radius

Given a set of $N$ elements that can form covalent single bonds in molecules, with $N(N-1)$ possible elemental pairings, it has proved possible to define a covalent radius for each element such that the actual bond length between any two elements that form a covalent single bond is roughly equal to the sums of their covalent radii.

### Creep

The time-dependent permanent deformation that occurs under stress; for most materials it is important only at elevated temperatures.

### Crosslinked Polymer

A polymer in which adjacent linear molecular chains are joined at various positions by covalent bonds.

### Cross-linking

A process forming chemical bonds between two separate molecular chains.

### Cross-sensitivity

The influence of one measurand on the sensitivity of a sensor, another measurand.

### Crosstalk

Electromagnetic noise transmitted between leads or circuits in close proximity to each other.

### Crybiology

The science of biology at low temperatures; research in cryobiology has made possible the freezing and storing of sperm and blood for later use.

### Cryoelectron Microscopy

Microscopy in which the samples are first stained immunocytochemically and then examined using an electron microscope. Immunoelectron microscopy is used extensively in diagnostic virology as part of very sensitive immunoassays.

### Cryogenic

Frozen at extremely low temperatures. The field of cryogenics is attempting to produce temperatures as close to absolute zero as possible. Absolute zero is the temperature at which molecules stop moving altogether.

### Crystal Grain Boundary

Crystalline materials are composed of atoms arranged in an orderly way, in precise relationships to one another.

A grain boundary is a narrow region of transition for one ordered area to another of different crystallographic orientation—an interface between one crystal grain and another.

### Crystal Structure

For crystalline materials, the manner in which atoms or ions are arrayed in space. It is defined in terms of the unit cell geometry and the atom positions within the unit cell.

### Crystallescence

In medical nanorobotics, the crystallization of solid solute that is offloaded by nanorobot sorting rotors at a concentration that exceeds the solvation capacity of the surrounding solvent.

### Curie Temperature

The temperature above which a ferromagnetic or ferrimagnetic material becomes paramagnetic. For iron the Curie point is 760oC and for nickel 356°C.

### Cycloaddition

A reaction in which two unsaturated molecules (or moieties within a molecule) join, forming a ring.

### Cyclotron

A type of particle accelerator in which an ion introduced at the center is accelerated in an expanding spiral path by use of alternating electrical fields in the presence of a magnetic field.

### Cytocarriage

In medical nanorobotics, the commandeering of a natural motile cell, by a medical nanorobot, for the purposes of in vivo transport, or to perform a herding function, or for other purposes.

### Cytopenetration

In medical nanorobotics, entry into cells by penetrating the plasma membrane.

## Cytoskeletolysis

In medical nanorobotics, purposeful destruction of the cellular cytoskeleton by a nanorobot, for cytocidal purposes.

## Cytovehicle

In medical nanorobotics, a living cell that has been commandeered by a medical nanorobot for use during cytocarriage.

# D

## Deep Space Network

A worldwide effort, coordinated by NASA, that communicates with spacecraft in an Earth orbit at any time in that spacecraft's orbital period. There are three ground stations in the Deep Space Network, each at a different place on the globe so that they can each reach a spacecraft when it is over a different position on the Earth.

The ground stations have very large dishes that act as *antennas* to receive and transmit signals to the spacecraft. There is also a control centre on Earth, run by NASA, that coordinates all the transmissions.

## Degenerate

Not a moral judgment but an adjective that describes multiple states that amount to the same thing: different triplet combinations of nucleotide bases (codons) that code for the same amino acid, for example.

## Degradation

A term used to describe the deteriorative processes that occur with polymeric materials, including swelling, dissolution, and chain scission. In medical nanorobotics, a crude form of functional navigation in which artificial conditions detectable by in vivo medical nanorobots are created at or near the target treatment site, such as warm or cold spots, pressure spots, or injected chemical plumes. Dramatic change in conformation of a protein or nucleic acid caused by heating or by exposure to chemicals and usually resulting in the loss of biological function.

## Denaturation

The breaking down of the

three-dimensional structure of a protein resulting in the loss of its function.

### Dendrimer

A dendrimer is an artificially manufactured or synthesized molecule built up from branched units called monomers. Such processes involve working on the scale of nanometers. Technically, a dendrimer is a polymer, which is a large molecule comprised of many smaller ones linked together.

### Dendrimers

From the Greek word dendra-tree, a dendrimer is polymer that branches. "...a tiny molecular structure that interacts with cells, enabling scientists to probe, diagnose, cure or manipulate them on a nanoscale." Invented by Professor Donald Tomalia from Central Michigan University.

### Dendrite

Extension of a nerve cell, typically branched and relatively short, that receives stimuli from other nerve cells

### Density Functional Theory

A branch of quantum theory of many electron systems. Density functional theory (DFT) rests on the two basic theorems by Kohn and Hohenberg which state that the ground state energy of the N-electron systems is a unique functional of the electron density distribution.

The theorems are quite important since they allow one to concentrate on the search for the electron density of the system, rather than the many-body quantum wave function, which is a much more complicated object. DFT had great success in explaining and calculating the electronic structure of molecules and solids, and is used extensively in chemistry and physics. DFT provides an information on the electron density distribution in a system of interest and its ground state energy. There are many variants of DFT, possibly the most promising ones being those dealing with time dependent implementation of DFT. Walter Kohn and John A. Pople received a Nobel Prize in chemistry (1998) for the development of density functional theory and computational methods in quantum chemistry.

### Deoxyribonucleic Acid

Deoxyribonucleic acid are long chains consisting of four kinds of nucleotides; the order of these nucleotides encodes the information needed to construct protein molecules. These in turn make up much of the molecular machinery of the cell. DNA is the genetic material of cells.

### Dermal Zippers

In medical nanorobotics, a theorized medical nanorobot that can rapidly perform incision-wound repairs to the dermis and epidermis; dermal zippers.

### Design Diversity

A form of redundancy in which components of different design serve the same purpose; this can enable systems to function properly despite design flaws.

### Desmosome

Specialized cell-cell junction, usually formed between two epithelial cells, characterized by dense plaques of protein into which intermediate filaments in the two adjoining cells insert.

### Diamagnetism

A weak form of induced or nonpermanent magnetism for which the magnetic susceptibility is negative. A type of magnetism associated with paired electrons, that causes a substance to be repelled from the inducing magnetic field.

### Diamondoid

Stuctures that resemble diamond in a broad sense, strong stiff structures containing dense, three dimensional networks of covalent bonds, formed chiefly from first and second row atoms with a valence of three or more. Many of the most useful diamondoid structures will in fact be rich in tetrahedrally coordinated carbon. Materials with superior strength to weight ratio, as much as 100 to 250 times as strong as Titanium, and much lighter. Possibly used to build stronger lighter rockets and space components, or a variety of other earth-bound articles for which weight and strength are a consideration.

### Dielectric (breakdown) Strength

The magnitude of an electric

field necessary to cause significant current passage through a dielectric material.

### Differential Labeling

When comparing the proteomes of two cell states (e.g. diseased vs. normal), gel- to-gel variability in spot position and protein yield often places the results of such experiments in question. Differential labeling enables one to analyze both states on a single gel, thus enabling direct comparison of protein levels. In this method, cells are treated with normal media, or media enriched in $^{15}N$. Corresponding proteins from each state will migrate to the same location on the gel, but analysis by mass spectrometry will distinguish the metabolically labeled peptides and thus quantify the two sets of proteins separately. This can have significant impact on reproducibility comparing experiments.

### Diffusion Coefficient

The constant of proportionality between the diffusion flux and the concentration gradient in Fick's first law. Its magnitude is indicative of the rate of atomic diffusion.

### Digital

Refers to systems employing only quantized (discrete) states to convey information (also see"analog").

### Diode

A diode is a specialized electronic component with two electrodes called the anode and the cathode. Most diodes are made with semiconductor materials such as silicon, germanium, or selenium. Diodes can be used as rectifiers, signal limiters, voltage regulators, switches, signal modulators, signal mixers, signal demodulators, and oscillators.

### Dip Pen Nanolithography

A direct-write soft lithography technique that is used to create nanostructures on a substrate of interest by delivering collections of molecules via capillary transport from an AFM tip to a surface.

### Dipole

A pair of equal yet opposite electrical charges that are separated by a small distance.

### Directed-Assembler

A specific type of assembler

that makes use of directed-assembly, such that the assembly process requires external energy or information input.

## Disassembler

In molecular nanotechnology, a nanomachine or system of nanomachines able to take an object apart while at each step recording the structure and composition of that object at the molecular level.

## Disasterbation

Idly fantasizing about possible catastrophes (ecological collapse, full-blown totalitarianism) without considering their likelihood or considering their possible solutions and preventions.

## Disease Models

Isolated models of disease can support the case for discovery or validation of a target but often do not accurately predict clinical efficacy of that target in humans. Moving all targets forward through development is prohibitive in terms of cost and time. Genomic approaches help rule out or invalidate a target while a good *in vivo* model can present a strong case for validation. However, modulation of many targets put forward for small molecule or biopharmaceutical development fails to reverse the disease phenotype in humans. *In vivo* delivery of siRNA provides the platform for high- value target validation. Down regulation of an individual gene is a powerful tool for understanding its biological function. Live animal imaging is a growing area that enables functional genomics and target validation.

## Disequilibration

In medical nanorobotics, maintenance or inducement of a state of perpetual ionic, chemical, or energetic disequilibrium in a living cell by a medical nanorobot, usually for the purpose of inducing cytocide.

## Dislocation

A linear crystalline defect around which there is atomic misalignment. Plastic deformation corresponds to the motion of dislocations in response to an applied shear stress. Edge, screw, and mixed dislocations are possible.

### Disruptive Technologies

Technology that is significantly cheaper than current, is much higher performing, has greater functionality, and is frequently more convenient to use. Will revolutionize markets by superseding existing technology. "Paradigm shifting" is a well-worn connotation. Although the term may sound negative to some, it is in fact neutral. It is only negative when businesses who are unprepared for change fail to adapt, only to fall behind and fail. The results are not *evolutionary*, they are *revolutionary*. In his book, *The Innovator's Dilemma*, Harvard professor Clayton Christensen used the term "disruptive technologies" to describe developments capable of displacing established market leaders.

### Dissolution

Deterioration in an organism such that its original structure cannot be determined from its current state.

### Distributed Intelligence

An intelligent entity which is distributed over a large volume (or inside another system, like a computer network) with no distinct center. This is the opposite to the strategy of Concentrated intelligences. Distributed intelligences have much longer communications lags, but are more flexible in their structure and can survive damage to their parts.

### DNA Diagnostics-miniaturization of

In the areas of sample preparation and assay, it is clear that miniaturization is key. To reduce the size of samples by a factor of 10 or greater, barriers in microfluidics, micromachining, robotics, microchemistry, nucleic acid chemistry, and surface chemistry must be overcome. To implement miniaturized protocols accurately and efficiently, substantial automation of the process will be required. In the development of miniaturized systems, it is essential that the system can be adapted for high levels of parallelization. Miniaturization poses significant technological risks. Currently, there exists no universally accepted precedent for the handling, replication, amplification, or cloning of DNA in nanoliter volumes. Due to the size and charge of the DNA molecule, and the relative instability of many of the enzymes involved in the sample preparation processes, nanoliter and less volumes may pose substantial challenges. In addition, interactions of the biological molecules with the surfaces of the reaction chambers must be minimized.

For some methodologies, it is not clear what the optimal sample will be, so substantial improvement in DNA fragmentation technologies or DNA cloning vectors may be required for the ultimate efficient application to diagnostics. Improvements in any of these areas are likely to be of value to other non- DNA based diagnostic applications such as antibody screening protocols and enzyme based diagnostics, because miniaturized robotic or micro- electro mechanical systems developed for DNA could be modified to be used for these purposes.

## DNA Probes

A DNA or nucleic acid probe is a short strand of DNA that locates and binds to its complementary sequence in samples containing single strands of DNA or RNA enabling identification of specific sequences. Nucleic acid probe assays exploit the fundamental hybridization reaction that occurs spontaneously between two complementary DNA:DNA or DNA:RNA strands.

As in immunoassays, detection of the hybrid requires that the probe be labeled. Various direct and indirect methods have been devised for the detection of the hybrid. Direct labeling involves attaching the label directly to the probe sequence; indirect labeling binds an antibody to the DNA:DNA or DNA:RNA hybrid. As in immunoassays, non-isotopically-labeled probes are preferred over radio-labeled probes primarily because of radiation hazards, disposal problems, and short reagent shelf life. In addition, the factors determining the detection limits of hybridization assays based on labeled probes are similar to those in immunoassays. Therefore, the development of a simple, inexpensive and sensitive direct detection system which eliminates the use of labels is highly desirable.

## DNA Sequencing

There are two main classical methods for sequencing DNA: The first method, developed by Allan Maxam and Walter Gilbert, involves chemicals used to cleave the DNA at certain positions, generating a set of fragments that differ by one nucleotide.

The second method, developed by Fred Sanger and Alan Coulson, involves enzymatic synthesis of DNA strands that terminate in a modified nucleotide. Analysis of fragments is similar for both methods and involves gel electrophoresis and autoradiography or fluorescence. The enzymatic method has largely replaced the chemical method as the technique of choice, although there are some situations where chemical sequencing can provide data more easily than the enzymatic method.

## DNA

A molecule encoding genetic information, found in the cell's nucleus. Deoxyribose Nucleic Acid. DNA is a molecule used in biological systems to store genetic information. It has a long, double stranded structure with the two strands wrapped around each other to form a right handed helix.

There are about 10 nucleotide pairs per helical turn. Each strand is composed of a sugar phosphate backbone and attached bases. It is connected to a complementary strand by hydrogen bonding (non- covalent) between paired bases, adenine (A) with thymine (T) and guanine (G) with cytosine (C). DNA is an abbreviation of DeoxyriboNucleic Acid. : It is the genetic 'blueprint' of all life on this planet. : DNA typing is often applied to Human Identification which has also been called DNA Fingerprinting. DNA Fingerprinting has been attributed to Alex Jeffries, who with his colleagues in 1985, discovered a sequence of nucleotides, subsequently termed short tandem repeats (STR) when comparing the DNA sequence of the human á-globin chromosomes isolated from different individuals.

## Dnhancer

Regulatory DNA sequence to which gene regulatory proteins bind, influencing the rate of transcription of a structural gene that can be many thousands of base pairs away.

### Domain

A region of a ferromagnetic or ferrimagnetic material in which all atomic or ionic magnetic moments are aligned in the same direction.

### Dopeyballs

Superconducting Buckyballs (they) have the highest critical temperature of any known organic compound.

### Doppler Effect

An apparent shift in the *frequency* of a wave. For example, when someone is listening to the sound of an ambulance siren, and that person is staying still but the ambulance is driving by, the person will hear a change in pitch of the ambulance siren. That change in pitch is caused by the doppler effect. The frequency of a sound wave determines the pitch, and the distance of the source of the sound from the sound's observer determines the amount that the frequency seems to have shifted, known as the doppler shift.

### Downlinked

The process by which a satellite sends information from space back to Earth. The satellite translates its computer information into radio waves and sends those waves back to Earth via its antenna. On Earth, another antenna, usually in the form of a dish, picks up the radio waves and translates them back into a form that computers can understand.

### DRAM

Dynamic Random Access Memory — memory in which each stored bit must be refreshed periodically.

### Drift

Gradual departure of the instrument output from the calibrated value. An undesired slow change of the output signal.

### Drug Design

Includes not only ligand design, but also pharmacokinetics (Pharmacogenomics) toxicity, which are mostly beyond the possibilities of structure- and/ or computer- aided design. Nevertheless, appropriate chemometric (Chemoinformatics) tools, including experimental design and multivariate statistics, can be of value in the planning and evaluation of pharmacokinetic and toxicological experiments and results. Drug design is most often used instead of the correct term "ligand design".

## Drug Discovery and Molecular Imaging

Molecular imaging's core role for the pharmaceutical industry is in drug discovery and development. Pure research is already profiting from cell-based molecular imaging, which will continue to be based on fluorescence, bioluminescence, and confocal microscopy. Applications in small animal imaging, lead characterization, and lead optimization are also discussed. The insights into basic cell biology that this research is yielding today will form the basis of drug development during the second half of the decade, as the results are absorbed by the pharmaceutical industry.

## Drug Discovery Desktop

No vendor provides a flexible, powerful, meaningful, and easy- to- use desktop. The paradigm of a desktop through which you access networks of information, organize your work, collaborate with your co-workers, and access your applications is prevalent in all of the industries, many of which have as much complexity as the drug discovery process. Yet, we find that previous generations of software have not been able to provide the flexibility enabling scientists to create their own drug discovery desktops. At the same time, an underlying structure is needed, allowing the corporation to maintain some measure of control over the process, reuse its knowledge assets, and retain its intellectual property.

## Drug Discovery

Pharmacologic therapeutics represent the single most important commercial product of biomedical research, and they have an unsurpassed track record of improving human health. Nevertheless, drug discovery and development is a costly, time consuming, and high risk activity. The process starts with the discovery of an agent or class of agents with particular activity. Lead compounds must then be identified, optimized, and only then tested in preclinical conditions for safety, toxicity, etc. Those agents still considered viable after such rigorous scrutiny are then brought to human subjects for clinical evaluation of a variety of aspects of the agent, including safety, effectiveness, and dosage determination.

## Drug Discovery-genomics Based

Has the potential to identify those target molecules (including genes and proteins that do not belong to the target families addressed by current drugs) that underlie disease processes themselves, as opposed to symptoms. Together with structural proteomics, advanced chemical technologies (e.g., combinatorial chemistry technologies that involve the creation of diverse small- molecule libraries, chemical genomics and chemogenomics), and high-throughput screening, genomics- based drug discovery thus has the potential to create drugs that can address large unmet medical needs. However, good methods of target identification and validation will be necessary to realize this potential.

## Drug Ontology

A three year project to develop a highly structured drug knowledge base. Unlike existing reference sources, it is intended solely for use by software applications.

## Drug Profiling

Biology has considerable experience with gene and protein-centered informatics, but chemistry is at an earlier stage of developing databases that are truly compound- centric. The historical paradigm of identifying and optimizing hits for potency, and then looking to evaluate and optimize for ADME and toxicity properties is quickly shifting to a more parallel approach that considers ADME/Tox properties at an earlier stage. This concept is epitomized by methods for differentiating between drug- like and non- drug- like compounds, the use of which is increasing significantly. Moving compound profiling earlier means that many more compounds must be assessed, which is both the value and the challenge of this shift.

### Drug Prototypes

Considered to be the first pure compound to have been discovered in any series of chemically or developmentally related therapeutic agents. A few prototypes have not been developed further because this has been unnecessary, commercially unacceptable or else unsuccessful. Some prototypes continue to serve as medicinal compounds in their own right, while others have been rendered obsolete by the analogues derived from them.

### Drug Selection

Traditionally the movement of compounds along the pipeline has been a fairly linear process. Because ways of speeding up the process can be enormously economically rewarding, greater attention is being paid to moving compounds along faster, trying to insure that compounds which will eventually fail, fail earlier, and looking at ways of revising the process to perform some evaluations in parallel rather than sequentially.

### Drug-like, Drug-likeness

Aqueous solubility and permeability data must be provided to chemistry as early as possible to avoid oral absorption problems. The minimum acceptable solubility for a drug depends on its permeability and projected clinical potency.

### Dry Nanotechnology

Derives from surface science and physical chemistry, focuses on fabrication of structures in carbon silicon, and other inorganic materials. Unlike the 'wet' technology, 'dry' techniques admit use of metals and semiconductors. The active conduction electrons of these materials make them too reactive to operate in a 'wet' environment, but these same electrons provide the physical properties that make 'dry' nanostructures promising as electronic, magnetic, and optical devices. Another objective is to develop 'dry' structures that possess some of the same attributes of the self-assembly that the wet ones exhibit.

### DSP

Digital Signal Processing; a process by which a sampled and

digitized data stream (real-time data such as sound or images) is modified in order to extract relevant information. Also, a digital signal processor.

## DumbSizing

Apealing to the least common denominator by explaining difficult concepts in such a manner so they loose meaning. Also, talking down to someone less informed or learned.

## Dynamic Characteristics

A description of an instrument's behavior between the time a measured quantity changes value and the time the instrument obtains a steady response.

## Dyson Scenario, the

Life expands into the universe, which is open. As the universe cools, life stores energy to survive (do information processing). It waits until the universe is cool enough, performs some processing with part of its energy stores, then waits until the universe has cooled so much that the remaining energy can be used to do an equal amount of computation, and so on. Essentially life has to adapt as the universe grows older, changing itself to be able to survive when the stars grow cold. If the universe is open, there will be plenty of time to work in, but energy will become very scarce. Dyson has shown that a finite amount of energy is enough to guarantee infinite survival if it is spent sufficiently slowly.

## Dyson Sphere

A shell built around a star to collect as much energy as possible, originally proposed by Freeman Dyson (although he admits to have borrowed the concept from Olaf Stapledon's novel *Star Maker* (1937)). In the original proposal the shell consists of many independent solar collectors and habitats in separate orbits (also known as a Type I Dyson Sphere), but later people have discussed rigid shells consisting of only one piece (called a Type II Dyson Sphere). The latter construction is unfortunately both unstable (since it will experience no net attraction of the star), requires super-strong materials and have no internal gravity. The Dyson Sphere is a classic example of mega-technology and common in Science Fiction.

## Dystopia

A dystopia (alternatively, cacotopia, kakotopia or anti-utopia) is a fictional society that is the antithesis of utopia.

A dystopia is usually characterized by an authoritarian or totalitarian form of government, or some other kind of oppressive social control. The first use of the word has been credited to John Stuart Mill in 1868, whose knowledge of Greek would suggest that he meant it as a place where things are bad, rather than simply the opposite of Utopia. The Greek prefix "dys" or "dis" signifies "ill", "bad" or "abnormal", whereas "ou" means "not" (Utopia means "nowhere", and is a pun on "Eutopia" meaning "happy place" - the prefix "eu" means "well"). So "dystopia" and "utopia" are not exact opposites in the sense that "dysphoria" and "euphoria" are opposites. The term "dystopia" itself is a combination of the Greek prefix "dys" and "topia" (from Greek, "topos" = "place"). "Dystopia," therefore, literally means "bad place." Sometimes referred to as a "Negative Utopia."

# E

## Ecophagy

Ecophagy is a term coined by Robert Freitas that means, literally, the consuming of an ecosystem. Freitas used the term to describe a scenario involving molecular nanotechnology gone awry. In this situation (called the grey goo scenario) out-of-control self-replicating nanorobots consume entire ecosystems, resulting in global ecophagy. However, the word "ecophagy" is now applied more generally in reference to any event—nuclear war, the spread of monoculture, massive species extinctions—that might fundamentally alter the planet. Scholars suggest that these events might result in ecocide in that they would undermine the capacity of the earth to repair itself. Others suggest that more mundane and less spectacular events—the unrelenting growth of the human population, the steady transformation of the natural world by human beings—will eventually result in a planet that is considerably less vibrant, and one that is, apart from humans, essentially lifeless. These people believe that the current human trajectory puts us on a path that will eventually lead to ecophagy.

## Effective Fuel Catalyst

In order for a catalyst to be effective when additised into fuel three principal properties are required. Firstly, complete hydrocarbon oxidation should be promoted, secondly formation of NOx should not be favoured, and finally the catalyst should remain thermally stable. As has been discussed, the ability of cerium oxide to donate oxygen allows complete hydrocarbon and soot burning in principle. In practice the activation energy of the cerium oxide, i.e. the lowest temperature at which oxygen donation occurs, is a crucial factor. Although the gas temperature of a diesel engine is high (approximately 1700°C), a low catalyst 'switch-on' temperature will clearly promote a more complete fuel burn in the milliseconds during which combustion occurs. A lower overall burn temperature, notwithstanding transient higher maximum temperatures, also mitigates against NOx production due to the high activation energy of nitrogen oxidation.

## El-Emergent Intelligence

An intelligent system that gradually emerges from simpler systems, instead of being designed top down.

## Elastomeric Stamp or Mould

Key element in soft lithography usually made from polydimethylsiloxane (PDMS), having patterned relief structures on its surface.

## Electric Force Microscopy

An a.c.voltage, U, $U=U_w \sin(w t)$ of radian frequency, w, is present between the electrically conductive tip of the microscope and the electrode at the back of the sample. Both electrostatic forces (so-colled "Maxwell stress ") and electromechanically induced forces act on the tip. An oscillating electric field causes thickness vibrations of the sample due to electromechanical effects (inverse piezoelectric effect and electrostriction).

## Electrical Bistability

A phenomenon in which an object exhibits two states of different conductivity at the same applied voltage.

## Electrical Breakdown

Condition in which, particularly with high electric field, a nominal insulator becomes electrically conducting.

## Electron Cloud

The electrons in an atom are attracted to the protons in the nucleus by the electromagnetic force. This force binds the electrons inside an electrostatic potential well surrounding the smaller nucleus, which means that an external source of energy is needed in order for the electron to escape. The closer an electron is to the nucleus, the greater the attractive force. Hence electrons bound near the center of the potential well require more energy to escape than those at the exterior.

Electrons, like other particles, have properties of both a particle and a wave. The electron cloud is a region inside the potential well where each electron forms a type of three-dimensional standing wave—a wave form that does not move relative to the nucleus. This behavior is defined by an atomic orbital, a mathematical function that characterises the probability that an electron will appear to be at a particular location when its position is measured. Only a discrete (or quantized) set of these orbitals exist around the nucleus, as other possible wave patterns will rapidly decay into a more stable form. Orbitals can have one or more ring or node structures, and they differ from each other in size, shape and orientation. Each atomic orbital corresponds to a particular energy level of the electron. The electron can change its state to a higher energy level by absorbing a photon with sufficient energy to boost it into the new quantum state. Likewise, through spontaneous emission, an electron in a higher energy state can drop to a lower energy state while radiating the excess energy as a photon. These characteristic energy values, defined by the differences in the energies of the quantum states, are responsible for atomic spectral lines.

The amount of energy needed to remove or add an electron (the electron binding energy) is far less than the binding energy of nucleons. For example, it requires only 13.6 eV to strip a ground-state electron from a hydrogen atom. Atoms are electrically neutral if they have an equal number of protons and electrons.

Atoms that have either a deficit or a surplus of electrons are called ions. Electrons that are farthest from the nucleus may be transferred to other nearby atoms or shared between atoms. By this mechanism, atoms are able to bond into molecules and other types of chemical compounds like ionic and covalent network crystals.

## Electrochemical Biosensor

A self- contained integrated device, which is capable of providing specific quantitative or semi- quantitative analytical information using a biological recognition element (biochemical receptor) which is retained in direct spatial contact with an electrochemical transduction element. Because of their ability to be repeatedly calibrated, we recommend that a biosensor should be clearly distinguished from a bioanalytical system, which requires additional processing steps, such as reagent addition. A device which is both disposable after one measurement, i.e., single use, and unable to monitor the analyte concentration continuously or after rapid and reproducible regeneration should be designated a single use biosensor.

## Electrochemical Gradient

Driving force that causes an ion to move across a membrane due to the combined influence of a difference in its concentration on the two sides of the membrane and the electrical charge difference across the membrane.

## Electrochemistry

The study of chemical changes resulting from electrical action and electrical activity resulting from chemical changes.

## Electroluminescence

In electrical engineering: the emission of visible light by a p-n junction across which a forward-biased voltage is applied. In electrochemistry: emission of light by a molecule which is being reduced or oxidized on a biased electrode. If the exciting cause is a photon, rather than an electron, the process is called photoluminescence.

## Electrolyte

Structures at the heart of a broad family of potentiometric silicon sensors. The best-known member of the family is the ion-sensitive field effect transistor, known as the ISFET or CHEMFET and the light-addressable potentiometric sensor LAPS . The principle of operation of devices using such structures is as follows. A potential with respect to a reference electrode is generated at the interface between the liquid solution and the insulator.

The surface potential ($ø_0$) is determined by that ionic species which has the fastest exchange rate ($i_o$) with the membrane covering the insulator. If no intentional membrane is deposited on an oxide covered insulator that species will be $H^+$. Surface potential changes in turn change the Si flat-band voltage $V_{FB}$. The flat-band voltage is the potential one needs to apply to the Si in order to have the bands flat throughout the semiconductor. The flat band voltage of an EIS structure has been shown to be given by: $V_{FB} = E_{REF} - Ö^{Si}/q - ø_0 - Q_{ins}/C_{inss}$ where $V_{FB}$ stands for the flat-band voltage of the structure, $E_{REF}$ for the reference electrode potential, $Ö^{Si}$ for the work function of silicon, $ø_0$ for the surface potential at the insulator/electrolyte interface, $Q_{ins}$ for the charge at the insulator/silicon interface and $C_{ins}$ for the insulator capacitance. At least two terms in the above equation are not known with a precision greater than a few hundred millivolts. This is true for $E_{REF}$ as well as for $Q_{ins}/C_{ins}$ which can vary from device to device by several hundred millivolts. For a given EIS sensor, these inaccurately known quantities are constant, and variations in flat-band voltage can be equated to variations of the surface potential.

## Electromagnetic Spectrum

Different kinds of electromagnetic waves can be classified by their wavelenghts. They are classified into sections called bands. The electromagnetic spectrum is the collection of these bands. The following types of waves make up the electromagnetic spectrum: gamma rays, x-rays, ultraviolet rays, visible light, infrared waves, and radio waves. The length of these waves ranges from $10^{-12}$ metres to $10^2$ metres long; this is known as the wavelength. Click here for a picture of the electromagnetic spectrum.

## Electromotive Force (emf) Series

A series of chemical elements arranged in order of their electromotive force. The electromotive force is the greatest potential difference that can be generated by a particular source of electric current. In practice this potential may be observable only when the source is not supplying current, because of its internal resistance.

## Electron Acceptor

Atom or molecule that takes up electrons readily, thereby gaining an electron and becoming reduced.

## Electron Carrier

Molecule such as cytochrome c that transfers an electron from a donor molecule to an acceptor molecule.

## Electron Donor

Molecule that easily gives up an electron, becoming oxidized in the process.

## Electron Microscopy

Visual and photographic microscopy in which electron beams with wavelengths thousands of times shorter than visible light are used in place of light, thereby allowing much greater magnification. In high-resolution electron microscopy one can begin to do "crystallography without crystals", averaging thousands of images of single molecules or other assemblies to reveal near atomic level structure. These methods demand intense computing hardware, software and algorithm development.

## Electron Transport Chain

Biomolecular machinery present in prokaryotic membranes and eukaryotic mitochnodria that couples the flow of electrons to proton pumps in order to convert energy from sugar to ATP.

## Electron Transport

Movement of electrons from a higher to a lower enery level along a series of electron carrier molecules, as in oxidative phosphorylation and photosynthesis.

## Electron

Electrons are negatively charged elementary particles with a mass of $9.1093897 \times 10^{-31}$ kg, a charge of $-1.60217733 \times 10^{-19}$ C, and spin 1/2. They cannot be divided into smaller particles. What is the radius of an electron? There is not an easy answer to this question. In an atom, the wavefunction of an electron has a width of about $10^{-10}$ meters. The wavefunction for an electron in a metal or travelling through vacuum can be much larger; it can be meters in width. Elementary negative particle whose charge is 1.602 $x10^{-19}$ coulombs.

## Electron-beam Lithography

An electron beam can be generated by extracting electrons from a sharp needle with an electric field. If the electrons are then accelerated towards a metal plate with a small hole in it, a narrow beam of electrons emerges from this hole. The electrons are typically accelerated through a potential of several kilovolts and are traveling at a good fraction of the speed of light. An electron beam can be deflected by electric and magnetic field which makes it possible to write with the beam. Such electron beams are used in televisions and computer monitors.

It is possible to focus an electron beam to have a very small diameter of about 1 nanometer. Such narrowly focused beams are used in scanning electron microscopes, transmission electron microscopes, and electron-beam pattern generators. Typically in an electron-beam pattern generator, the electron beam writes a pattern in a thin organic film that is coated over a wafer. Usually the wafer is a single crystal of silicon that is about 0.5 mm thick and 10-30 cm in diameter. The thin (~100 nm) organic film contains long molecules that wrap around each other like cooked spagetti. The energetic electrons in the electron beam cut these molecules up into small pieces. The film is then dipped in developer which disolves the short pieces but leaves the long unexposed sections unaffected. A pattern is thereby defined in the resist.

## Electronegative

Describing elements that tend to gain electrons and form negative ions. The halogens are typical electronegative elements.

## Electronic Nose

An "electronic or artificial nose" is an instrument, which comprises a sampling system, an array of chemical gas sensors with differing selectivity, and a computer with an appropriate pattern- classification algorithm, capable of qualitative and/or quantitative analysis of simple or complex gases, vapors, or odors.

### Electronic Tongues

For liquid analysis, based on the organizational principles of biological sensory systems, developed rapidly during the last decade. ... The exciting possibility of establishing a correlation between the output from an electronic tongue and human sensory assessment of food flavour, thereby enabling quantification of taste and flavour, is described.

Application areas of electronic tongue systems including foodstuffs, clinical, industrial, and environmental analysis are discussed in depth.

### Electronvolt

An electronvolt is defined as $1.6022 \times 10^{-19}$ Joules. The energy needed to move an electron from a metal electrode at voltage $V_1$ to an electrode at voltage $V_2$ is $V_1$-$V_2$ electron volts. One electronvolt is about the amount of energy needed to dissociate a molecule.

### Electrostatic Flocking

A process using an electrostatic charge to drive flock fibers into an adhesive that has been printed on a substrate.

### Elliptical Orbit

An orbit where the satellite moves in a big ellipse (a shape like a flattened circle). A satellite in an elliptical orbit is sometimes close to the object it is orbiting and sometimes far away from it. A satellite in an elliptical orbit, especially a highly eccentric one, will move faster when it is close to the planet and slower when it is far away. An elliptical orbit can be useful to a communications satellite.

### Elongation Factor

Protein required for the addition of amino acids to growing polypeptide chains on ribosomes.

### Emergence

A complex whole created by simple parts, as in the brain where billions of neurons work individually, but collectively make up our consciousness and give us the ability to think, rationalize, and create.

### Emulation

An absolutely precise simulation of something, so exact that it is equivalent to the original (for example, many computers emulate obsolete computers to run their programs).

## Enabling Science and Technologies

Areas of research relevant to a particular goal, such as nanotechnology. Also, technology that "enables" other technology to advance, such as the transistor enabled the computer chip revolution, as did photolithography.

## Encryption Device

An instrument often used in spy satellites. It encodes digital images or data before it is sent back to Earth.

## Endocrine Cell

Specialized animal cell that secretes a hormone into the blood; usually part of a gland, such as the thyroid or pituitary gland.

## Endocytosis

Uptake of material into a cell by an invagination of the plasma membrane and its internalization in a membrane-bounded vesicle.

## Endoplasmic Reticulum

Labyrinthine, membrane-bounded compartment in the cytoplasm of eucaryotic cells, where lipids are synthesized and membrane-bound proteins are made.

## Endosome

Membrane-bounded organelle in animal cells that carries materials newly ingested by endocytosis and passes many of them on to lysosomes for degradation.

## Endothelium

Single sheet of highly flattened cells (endothelial cells) that forms the lining of all blood vessels. Regulates exchanges between the bloodstream and surrounding tissues and is usually surrounded by a basal lamina.

## Energy Levels

When an electron is bound to an atom, it has a potential energy that is inversely proportional to its distance from the nucleus. This is measured by the amount of energy needed to unbind the electron from the atom, and is usually given in units of electronvolts (eV). In the quantum mechanical model, a bound electron can only occupy a set of states centered on the nucleus, and each state corresponds to a specific energy level. The lowest energy state of a bound electron is called the ground state, while an electron at a higher energy level is in an excited state. In order for an electron to transition between two different states, it must absorb or emit a photon at an energy matching the difference in the potential energy of those levels. The energy of an emitted photon is proportional to its frequency, so these specific energy levels appear as distinct bands in the electromagnetic spectrum. Each element has a characteristic spectrum that can depend on the nuclear charge, subshells filled by electrons, the electromagnetic interactions between the electrons and other factors.

When a continuous spectrum of energy is passed through a gas or plasma, some of the photons are absorbed by atoms, causing electrons to change their energy level. Those excited electrons that remain bound to their atom will spontaneously emit this energy as a photon, traveling in a random direction, and so drop back to lower energy levels. Thus the atoms behave like a filter that forms a series of dark absorption bands in the energy output. (An observer viewing the atoms from a different direction, which does not include the continuous spectrum in the background, will instead see a series of emission lines from the photons emitted by the atoms.) Spectroscopic measurements of the strength and width of spectral lines allow the composition and physical properties of a substance to be determined.

Close examination of the spectral lines reveals that some display a fine structure splitting. This occurs because of spin-orbit coupling, which is an interaction between the spin and motion of the outermost electron. When an atom is in an external magnetic field, spectral lines become split into three or more components; a phenomenon called the Zeeman effect. This is caused by the interaction of the magnetic field with the magnetic moment of the atom and its electrons. Some atoms can have multiple electron configurations with the same energy level, which thus appear as a single spectral line. The interaction of the magnetic field with the atom shifts these electron configurations to slightly different energy levels, resulting in multiple spectral lines. The presence of an external electric field can cause a comparable splitting and shifting of spectral lines by modifying the electron energy levels, a phenomenon called the Stark effect.

If a bound electron is in an excited state, an interacting photon with the proper energy can cause stimulated emission of a photon with a matching energy level. For this to occur, the electron must drop to a lower energy state that has an energy difference matching the energy of the interacting photon. The emitted photon and the interacting photon will then move off in parallel and with matching phases. That is, the wave patterns of the two photons will be synchronized. This physical property is used to make lasers, which can emit a coherent beam of light energy in a narrow frequency band.

## Engulf Formation

In medical nanorobotics, a configuration that may be adopted by a metamorphic nanorobot, in which the nanorobot reshapes itself to create an interior cavity capable of trapping a living cell, virion, or other biological particle.

## Enhancer

Regulatory DNA sequence to which gene regulatory proteins bind, influencing the rate of transcription of a structural gene that can be many thousands of base pairs away.

## Entanglement

From quantum mechanics, entanglement is a relationship between two systems in which they exist in more than one state simultaneously (by superposition), and the state of one system determines the state of the other.

## Enternal Computer

The satellite's method of storing and analyzing data collected by the satellite, and a way of controlling its various systems. The technical term for this satellite subsystem is *TT&C* (Telemetry, Tracking, and Control).

## Enthalpy (H)

A property of a system equal to E + PV, where E is the internal energy of the system, P is the pressure of the system, and V is the volume of the system. At constant pressure the change in enthalpy equals the energy flow as heat.

## Entropy

In chemistry, physics and thermodynamics, thermodynamic entropy, symbolized by *S*, is a differential term "dQ/T", where *dQ* is the amount of heat absorbed reversibly by a thermodynamic system at a temperature *T*. German physicist Rudolf Clausius introduced the mathematical concept of entropy in the early 1850s to account for the dissipation of energy in thermodynamic systems that produce work. He coined the term based on the Greek *ÄÁ¿À•* meaning "transformation". Although the concept of entropy was originally a thermodynamic construct, it has since been adapted and applied to many disparate fields of study, including statistical mechanics, thermal physics, statistics, information theory, psychodynamics, economics, and evolution. A thermodynamical quantity related to heat and temperature. Carries an information on the amount of disorder (loosely speaking) in the system and its ability to do useful work—the greater the disorder, the smaller ability of the system to do work. The entropy of the system in a given state is related to a number of microscopic arrangements of the system constituents (particles) which yield the same macroscopic state with given thermodynamical parameters (temperature, volume, pressure). According to second law of thermodynamics, the entropy of a closed system can only increase to reach its maximum, or remain constant.

This means that the system becomes more 'disordered' as time progresses and less useful for extracting work from it. Open systems (such as organisms) can become more 'ordered' with time due to the fact that the 'negative' entropy is 'pumped' in them from their surrounding (the surrounding becomes more disordered), but the entropy of the whole Universe still increases. This basic law of physics puts some doubts on the possibility of making nanoscopic machines, since it would require positioning of the molecules in very improbable configurations. Such structures would have a very low entropy and would thus tend to degrade (disorder) unless they would be able to extract energy and negative entropy from their surrounding which would then enable them to preserve their structure and purpose.

## Enzyme immunoassay/EIA

In an enzyme immunoassay (EIA), an enzyme-labeled antibody or antigen is used for the detection and quantification of the antigen-antibody reaction. In an electrochemical EIA, the enzyme-catalyzed reaction is monitored electrochemically (amperometric, potentiometric, voltametric or conductometric). In EIA, the antibody-antigen reaction furnishes the needed specificity. The enzyme label provides the sensitivity via chemical amplification. A large molecule, usually a protein, that catalyzes biological reactions. A protein that acts as a catalyst in a biochemical reaction. Protein that catalyzes a specific chemical reaction.

## Equator

The equator is an imaginary circle drawn around a planet (or other astronomical object) at a distance halfway between the poles. The equator divides the planet into a Northern Hemisphere and a Southern Hemisphere. The latitude of the equator is, by definition, 0°. The length of Earth's equator is about 40,075.0 km, or 24,901.5 miles.

The equator is one of the five main circles of latitude based on the relationship of the Earth's rotation and plane of orbit around the sun. Additionally, the equator is the only line of latitude which is also a great circle.

The Sun, in its seasonal movement through the sky, passes directly over the equator twice each year on the March and September Equinoxes. At the equator, the rays of the sun are perpendicular to the surface of the earth on these dates. Places near the equator experience the quickest rates of sunrise and sunset in the world, taking minutes. Such places also have a relatively constant amount of day/night time on every day throughout the year compared with more northerly or southerly places. In tourist areas, the equator is often marked on the sides of roads Also, in relation to launching of satellites, the closeness of the launch centre requires the thrust to be comparatively lesser, and as one moves away from the equator, one has to ensure that the payload in the launch vehicle is not much, for to get into the right orbit, more thrust is required. In many tropical regions people identify 2 seasons, wet and dry, but most places very close to the equator are wet throughout the year, although seasons can vary depending on a variety of factors including elevation and proximity to an ocean. The surface of the Earth at the equator is mainly ocean. The highest point on the Equator is 4,690 m, at 77°592 31" W on the south slopes of Volcán Cayambe (summit 5,790 m) in Ecuador. This is a short distance above the snow line, and is the only point on the Equator where snow lies on the ground

## Equatorial Orbit

The satellite flies along the line of the Earth's equator. Equatorial orbits can be useful for satellites observing tropical weather patterns, as they can monitor cloud conditions around the globe.

## Error Signal Mode

Error signal mode produces an image that approximates the derivative of C-AFM topographic images, yielding high image contrast and enhancing surface and sub-surface details.

## Exploratory Engineering

Design and analysis of systems that are theoretically possible but cannot be built yet, owing to limitations in available tools.

# F

### Faint Object Camera

The Faint Object Camera (FOC) was a camera installed on the Hubble Space Telescope until 2002. It was replaced by the Advanced Camera for Surveys.

The camera was built by Dornier GmbH and was funded by the European Space Agency. The unit actually consists of two complete and independent camera systems designed to provide extremely high resolution, exceeding 0.05 arcseconds. It is designed to view very faint UV light from 115 to 650 nanometers in wavelength.

### Farad

The farad (symbol: F) is the SI unit of capacitance.

A capacitor has a value of one farad when one coulomb of stored charge causes a potential difference of one volt across its terminals. Its equivalent expression in SI base units..

### Faraday

In physics, the faraday (not to be confused with the farad) is a unit of electrical charge; one faraday is equal to the charge of $6.02 \times 10^{23}$ electrons (one mole). The faraday is no longer in general use and has been replaced by the SI unit coulomb; one faraday is approximately equivalent to 96485.3415 coulombs.

### Fatty Acid

Compound such as palmitic acid that has a carboxylic acid attached to a long hydrocarbon chain. Used as a major source of energy during metabolism and as a starting point for the synthesis of phospholipids.

### Fault-tolerant

Describes a system that can suffer failure in a component or subsystem, yet continue to function correctly.

### Femtoengineering

Will involve engineering using mechanisms within a quark.

### Femtosecond

Femtosecond is one quadrillionth of a second, and is to a second what a second is to 32,700,000 years. At 186,000 miles per second, in one femtosecond light travels only far enough to traverse about 1,000 silicon atoms.

When used to time a laser pulse, it allows for ultra-precise micromachining, with virtually no damage to surrounding material.

### Femtotechnology

Femtotechnology the art of manipulating materials on the scale of elementary particles (leptons, hadrons, and quarks). The next step smaller after picotechnology, which is the next step smaller after nano-technology.

### Fermi Energy

The energy level in a solid at which the probability of finding an electron is 1/2. For a metal, the energy corresponding to the highest filled electron state in the valence band at 0 K.

### Fermi Paradox

"If there are other intelligent beings in the Universe, why aren't they here?" Since it appears to be quite possible for a technological species to spread across the galaxy in less than 10 million years (using von Neumann machines) or otherwise change things on such a large scale that it would be very visible, the lack of such evidence is puzzling or implies that other technological civilizations doen't exist. There have been many attempts to explain this, for example the "Wildlife Preserve" idea (the aliens doesn't want to interfere with younger civilizations), that they transcend and become incomprehensible, that they hide or that they are actually here, hidden on the nanoscale, but the problem with these attempts is that most of them just explain why some aliens would not be apparent.

### Ferroelectric Material

A dielectric material such as Rochelle salt and barium titanate with a domain structure containing dipoles (asymmetric distributions of electrical charge) which spontaneously align. Their domain structure makes them analogous to ferromagnetic materials. They exhibit hysteresis and usually the piezoelectric effect.

### Ferromagnetism

Permanent and large magnetizations found in some metals (e.g. Fe, Ni and Co), resulting from the parallel alignment of neighboring magnetic moments.

### Fertilization

Fusion of a male and a female gamete (both haploid) to form a diploid zygote, which develops into a new individual.

### Fiber-optic

Relates to transmission of information as modulated light in tiny transparent fibers instead of copper wires.

### Fibrin Sealant

A biological tissue glue derived from human plasma. Sometimes referred to as fibrin glue, it contains the components necessary for the last step of coagulation: i) fibrinogen and other coagulation factors and ii) thrombin. It is used in surgical procedures to arrest bleeding and as an adjunct to wound healing.

### Fibrinogen

Plasma glycoprotein clotted by thrombin, composed of a dimer of three non- identical pairs of polypeptide chains (alpha, beta, gamma) held together by disulfide bonds. Fibrinogen clotting is a sol- gel change involving complex molecular arrangements: whereas fibrinogen is cleaved by thrombin to form polypeptides A and B, the proteolytic action of other enzymes yields different fibrinogen degradation products.

A plasma protein that, when activated by thrombin, becomes fibrin, the principle component of a blood clot. Plasma glycoprotein clotted by *thrombin*, composed of a dimer of three non- identical pairs of polypeptide chains (alpha, beta, gamma) held together by disulfide bonds. Fibrinogen clotting is a sol- gel change involving complex molecular arrangements: whereas fibrinogen is cleaved by thrombin to form *polypeptides* A and B, the proteolytic action of other enzymes yields different fibrinogen degradation products.

### Fibroblast

Common cell type found in connective tissue. Secretes an extracellular matrix rich in collagen and other extracellular matrix macromolecules. Migrates and proliferates readily in wounded tissue and in tissue culture.

### Fick's First Law

The diffusion flux is proportional to the concentration gradient. This relationship is employed for steady-state diffusion situations. The time rate of change of concentration is proportional to the second derivative of concentration. This relationship is employed in non-steady-state diffusion situations.

### Flagellum (plural flagella)

Long, whiplike protrusion whose undulations drive a cell through a fluid medium. Eucaryotic flagella are longer versions of cilia; bacterial flagella are completely different, being smaller and simpler in construction.

### Fluidic Self Assembly

A novel technique for accurately assembling large numbers of very small devices. The small size, planarity, and accuracy of the assembly also result in very low parasitic interconnects, comparable to on die traces. This massively parallel assembly process combines the capability and flexibility of assembly with the cost effectiveness of integration. Invented by Mr. Mark Hadley and was part of his Ph.D. dissertation while he was studying at University of California, Berkley. The FSA process became the foundation for the origins of a new company named Alien Technology Corporation.

In the FSA process, specifically shaped semiconductor devices ranging in size from 10 microns to several hundred microns are suspended in liquid and flowed over a surface which has correspondingly shaped "holes" or receptors on it and into which the devices settle.

The shape of the devices and of the holes is designed so that the devices fall easily into place and are selfaligning. Alien has successfully demonstrated the assembly of tens of thousands of devices in a single process step.

## Fluidic System

Device for synthesis or screening in which fluids such as reagents or assay buffers may be directed to specified locations by the opening and closing of valves in a stationary network of tubes and wells.

## Fluorescence Microscopy

Microscopy of specimens stained with fluorescent dye (usually fluorescein isothiocyanate) or of naturally fluorescent materials, which emit light when exposed to ultraviolet or blue light. Immunofluorescence microscopy utilizes antibodies that are labeled with fluorescent dye.

## Fluorescence Polarization

Based on the observation that emission signals from small fluorescent molecules are relatively depolarized, while binding to a larger molecule reduces the tumbling rate of the fluorescer, resulting in a relatively polarized emission signal.

First described in 1926 (Perrin) and has been a powerful tool in the study of molecular interactions. When fluorescent molecules are excited with plane polarized light, they emit light in the same polarized plane, provided that the molecule remains stationary throughout the excited state (4 nanoseconds in the case of fluorescein). However, if the excited molecule rotates or tumbles out of the plane of polarized light during the excited state, then light is emitted in a different plane from that of the initial excitation.

If vertically polarized light is used to excite the fluorophore, the emission light intensity can be monitored in both the original vertical plane and also the horizontal plane.

The degree to which the emission intensity moves from the vertical to horizontal plane is related to the mobility of the fluorescently labeled molecule. If fluorescently labeled molecules are very large, they move very little during the excited state interval, and the emitted light remains highly polarized with respect to the excitation plane.

### Fluorescence

Luminescence which persists less than a second after the exciting cause has been removed. If the luminescence persists significantly longer it is called phosphorescence.

### Fluorescent Dye

Molecule that absorbs light at one wavelength and responds by emitting light at another wavelength; the emitted light is of longer wavelength (and hence of lower energy) than the light absorbed.

### Focal Contact

Small region on the surface of a fibroblast or other cell that is anchored to the extracellular matrix. The attachment is mediated by transmembrane proteins such as integrins, which are linked, through other proteins, to actin filaments in the cytoplasm.

### Focal Length

The distance needed between a lens and an object to make the object visible and in focus. The human eye has different focal lengths depending on how much the muscles in the eye around the lens are contracted. That is how the human eye focuses.

### Footprint

The area with which a satellite in *geostationary orbit* can communicate. A footprint can be as large as an entire country; for example, many Canadian satellites have footprints almost the entire size of Canada, from coast to coast. Click here for a picture of a satellite's footprint.

### Formulation

Advancing technologies and increased regulatory scrutiny throughout drug development, have heightened interest in drug formulation and delivery. Applications of methods and technologies to characterize and optimize the physical and chemical properties of drugs. Special attention will be given to overcoming hurdles associated with crystallization, polymorphism and disordered drug delivery.

## Forward Bias

The conducting bias for a p-n junction rectifier that assures electron flow to the n side of the junction.

## Fractal Mechatronic Universal Assembler

(or Fractal Assembler) is a machine that is capable of assembling any chemical from a generic descriptions of the properties required of the chemical. The machine comprises of test tube arrays and software linked to robotic cubes and sensor arrays to implement automated mixing and testing to conduct materials research activity.

## Fractal Robots

Fractal Shape Shifting Robots and Programmable "Digital Matter", are programmable machines that can do unlimited tasks in the physical world, the world of matter. Load the right software and the same "machines" can vacuum the carpet, paint your car, or construct an office building and later, wash that building's windows. This is the beginning of "Digital Matter". Fractal Shape Shifting Robots look like "Rubic's Cubes" that can "slide" over each other on command, changing and moving in any overall shape desired for a particular task. These cubes communicate with each other and share power through simple internal induction coils (or surface contacts in some models), have batteries, a small computer and various kinds of internal magnetic and electric inductive motors (depending on size) used to move over other cubes. The ultimate goal is self sustaining systems and "self-assembly" features that can drop cost dramatically and enable successive generations of robots exhibiting greater utility and value, to be built along the way.

## Free Energy

Energy that can be extracted from a system to drive reactions. Takes into account changes in both energy and entropy.

## Free Radicial

A molecule containing an unpaired electron, typically highly unstable and reactive. Free radicals can damage the molecular machinery of biological systems, leading to cross-linking and mutation.

### Free-energy Change

Change in the free energy during a reaction: the free energy of the product molecules minus the free energy of the starting molecules. A large negative value of DG indicates that the reaction has a strong tendency to occur.

### Frequency Bands

The grouping together of electromagnetic waves of a similar frequency. In spectroscopic studies, one type of wave from the electromagnetic spectrum will show up as a band of colour that can be interpreted as a frequency band. This allows scientists to figure out what kind of energy is coming from a particular source For example, X-rays would make up a frequency band. Some bands are wider than others.

### Frequency Response

Two relations between sets of inputs and outputs. One relates frequencies to the output-input amplitude ratio; the other relates frequencies to the phase difference between the output and input.

### Frequency

Frequency is the measurement of the number of times that a repeated event occurs per unit of time. It is also defined as the rate of change of phase of a sinusoidal waveform.

To calculate the frequency of an event, the number of occurrences of the event within a fixed time interval are counted, and then divided by the length of the time interval.

In SI units, the result is measured in hertz (Hz), named after the German physicist Heinrich Rudolf Hertz. 1 Hz means that an event repeats once per second, 2 Hz is twice per second, and so on. This unit was originally called a cycle per second (cps), which is still used sometimes. Other units that are used to measure frequency include revolutions per minute (rpm) and radians per second (rad/s). Heart rate and musical tempo are measured in beats per minute (BPM).

# G

## Gastrula

The gastrula phase of embryonic development is seen in all animals except the sponges. It follows the blastula phase. The purpose of gastrulation is to position the three embryonic germ layers, the endoderm, ectoderm and mesoderm. During gastrulation, embryonic cells migrate through an opening within the embryo known as a blastocoel.

There are five main types of cell movements in gastrulation - ingression, involution, invagination, delamination and epiboly. Ingression is the movement of single cells inwards, involution is the inturning of a lower cell layer caused by movement of the upper layer, invagination is an infolding, or poking, of cells, delamination is when one sheet of cells split into two, and epiboly is when the embryo is encompassed by the ectoderm. In addition to these movements, Convergent Extension can also take place. Although it is not real movement it does allow the cells to stretch (shorter, longer, or taller). Once gastrulation is complete, all germ layers are in the correct location and further growth and organogenesis begins.

## Gene

Genes are the units of heredity in living organisms. They are encoded in the organism's genetic material (usually DNA or RNA), and control the physical development and behavior of the organism. During reproduction, the genetic material is passed on from the parent(s) to the offspring. Genetic material can also be passed between un-related individuals (e.g. via transfection, or on viruses). Genes encode the information necessary to construct the chemicals (proteins etc.) needed for the organism to function.

The word "gene" was coined in 1909 by Danish botanist Wilhelm Johannsen for the fundamental physical and functional unit of heredity. The word gene was derived from Hugo De Vries' term pangen, itself a derivative of the word pangenesis which Darwin (1868) had coined. The word pangenesis is made from the Greek words *pan* (a prefix meaning "whole", "encompassing") and *genesis* ("birth") or *genos* ("origin").

The term "gene" is shared by many disciplines, including classical genetics, molecular genetics, evolutionary biology and population genetics. Because each discipline models the biology of life differently, the usage of the word gene varies between disciplines. It may refer to either material or conceptual entities.

Following the discovery that DNA is the genetic material, the growth of biotechnology, and the project to sequence the human genome, the common usage of the word "gene" has increasingly reflected its meaning in molecular biology, namely the segments of DNA which cells transcribe into RNA and translate, at least in part, into proteins. The Sequence Ontology project defines a gene as: "A locatable region of genomic sequence, corresponding to a unit of inheritance, which is associated with regulatory regions, transcribed regions and/or other functional sequence regions".

In common speech, "gene" is often used to refer to the hereditary cause of a trait, disease or condition—as in "the gene for obesity." Speaking more precisely, a biologist might refer to an allele or a mutation that *has been implicated in* or *is associated with* obesity. This is because biologists know that many factors other than genes decide whether a person is obese or not: eating habits, exercise, prenatal environment, upbringing, culture and the availability of food, for example. Moreover, it is very unlikely that variations within a single gene—or single genetic locus—fully determine one's genetic predisposition for obesity.

These aspects of inheritance—the interplay between genes and environment, the influence of many genes—appear to be the norm with regard to many and perhaps most ("complex" or "multi-factoral") traits. The term phenotype refers to the characteristics that result from this interplay.

## General-purpose Molecular Manufacturing

A manufacturing technology that will find many applications across many segments of society. Its extreme flexibility, precision, high capacity, and low cost will cause rapid adoption almost everywhere, and therefore will have disruptive effects in many industries.

## Genetic Algorithm

Any algorithm which seeks to solve a problem by considering numerous possibilities at once, ranking them according to some standard of fitness, and then combining ("breeding") the fittest in some way. In other words, any algorithm which imitates natural selection.

## Genetic Engineering

Directed modification of the gene complement of a living organism by such techniques as altering the DNA, substituting genetic material by means of a virus, transplanting whole nuclei, transplanting cell hybrids, etc.

## Genomic Arrays

Allow toxicologists to look at cellular behavior in a completely new light. In a sense, recording individual gene responses to powerful insults such as alkylating agents was akin to studying the effects of poverty by monitoring a person's bank account—the complete picture is much larger than what is actually being measured. But genomic arrays simultaneously report indicators of multiple dimensions of the cellular response to stimuli. Now, in addition to gaining insight into basic cellular mechanisms of repair, researchers looking at a variety of indicators and responses of toxicity may gain some predictive power regarding individual compounds—and individual humans. Both academic and private laboratories have already begun work on finding genes that induce protection or sensitivity to toxicants in individual cells and people.

## Genomic Technologies

One of the primary reasons for the success of the Human Genome Project has been the development and use of high-throughput strategies for data generation, and the placement of the data immediately in the public domain. Most of the sequence data, the underlying maps and the sequence assemblies were generated through the use of large- scale automated processes.

Now, methods such as sequence analysis of whole genomes, DNA microarray technology and mass spectrometry have been or are being developed as high- throughput approaches for additional types of genomic analyses, such as determining the parameters of gene expression or the location of gene products by the thousands at a time instead of individually. High- throughput methods to determine the location of cis-regulatory elements and, to a lesser extent, other sequence elements, are also beginning to be developed.

However, at present, there is no single approach or compilation of approaches that can accurately and efficiently identify every sequence feature in genomic DNA.

## Geodesic Dome

A triangulation of a Platonic solid or other polyhedron to produce a close approximation to a sphere (Mathworld definition). Geodesic domes can be obtained iteratively by triangulating the faces of a Platonic solid (cube, dodecahedron, icosahedron, octahedron, and tetrahedron), and projecting the newly obtained vertices onto a circumsphere of a particular solid. Interesting structures (perhaps variants of geodesic domes) can be also obtained by designing specific 'nets' of points on each of the faces of the Platonic solid, not necessarily triangular.

## Geomagnetic

Having to do with the magnetic properties of the Earth. Geomagnetic activity includes anything that changes the numbers of charged particles found in one area of the Earth.

## Geostationary Orbit

An orbit in which a satellite appears to remain in the same spot in the sky all the time. When a satellite is in geostationary orbit, it travels at exactly the same speed as the Earth is rotating below it. A satellite in geostationary orbit is very high up, at 35 850 km above the Earth. Geostationary satellites are always located directly above the equator. The area with which a satellite in geostationary orbit can communicate is called its *footprint*.

## Giant Magnetoresistance (GMR)

It results from subtle electron-spin effects in ultra-thin 'multilayers' of magnetic materials, which cause huge changes in their electrical resistance when a magnetic field is applied. GMR is 200 times stronger than ordinary magnetoresistance.

## Gibbs Free Energy

The Gibbs free energy is the Helmholtz free energy plus the product of the system volume and the external pressure. Changes in the Gibbs free energy at a constant pressure thus include work done against external pressure as a system undergoes volumetric changes.

This energy proves convenient for describing equilibria in gases and liquids at a constant pressure (e.g., at one atmosphere), but is of little use in describing machine-phase chemical processes. Changes in the Gibbs free energy caused by a change in the applied pressure (at constant volume) have no direct physical significance.

## Global Positioning System

A satellite technology that uses mathematics to calculate the position in three dimensions (latitude, longitude, and altitude) of something on the Earth by measuring the time it takes for the satellite's radio transmissions, travelling at the speed of light, to reach the a receiver on the ground. It requires a fleet of satellites in space. Applications of this technology include determining a position on the Earth, measuring the Earth's movement after an earthquake, or locating drop points for airlifted relief supplies.

## Golden Goo

Another member of the grey goo family of nanotechnology

disaster scenarios. The idea is to use nanomachines to filter gold from seawater. If this process got out of control we would get piles of golden goo (the "Wizard's Apprentice Problem"). This scenario demonstrates the need of keeping populations of self-replicating machines under control; it is much more likely than grey goo, but also more manageable.

## Grain Boundary

The interface separating two adjoining grains having different crystallographic orientations.

## Grain Growth

The increase in average grain size of a polycrystalline material: for most materials, an elevated temperature heat treatment is necessary.

## Graininess

An effect seen in the print as randomly occurring light and dark specks or grains, due to roughing of the edges of halftone dots, random specks of ink between dots, discontinuous ink films, or specular reflectance off inked fibers in the surface of the paper.

## Gray Goo Problem

Nanotech Now's Goo and Paste Glossary. The following quote was taken from Eric Drexler's book Engines of creation. The early transistorized computers soon beat the most advanced vacuum-tube computers because they were based on superior devices. For the same reason, early assembler-based replicators could beat the most advanced modern organisms. "Plants" with "leaves" no more efficient than today's solar cells could out-compete real plants, crowding the biosphere with an inedible foliage. Tough, omnivorous "bacteria" could out-compete real bacteria: they could spread like blowing pollen, replicate swiftly, and reduce the biosphere to dust in a matter of days.

Dangerous replicators could easily be too tough, small, and rapidly spreading to stop—at least if we made no preparation. We have trouble enough controlling viruses and fruit flies. Among the cognoscenti of nanotechnology, this threat has become known as the "gray goo problem." Though masses of uncontrolled replicators need not be gray or gooey, the term "gray goo" emphasizes that replicators able to obliterate life might be less inspiring than a single species of crabgrass.

They might be "superior" in an evolutionary sense, but this need not make them valuable. We have evolved to love a world rich in living things, ideas, and diversity, so there is no reason to value gray goo merely because it could spread.

Indeed, if we prevent it we will thereby prove our evolutionary superiority.

The gray goo threat makes one thing perfectly clear: we cannot afford certain kinds of accidents with replicating assemblers.

### Green Ceramic Body

A ceramic piece, formed as a particulate aggregate, that has been dried but not fired.

### Green Goo

Nanomachines or bio-engineered organisms used for population control of humans, either by governments or eco-terrorist groups. Would most probably work by sterilizing people through otherwise harmless infections.

### Greenwich Meridian

The central *meridian* line from which all time zones are set.

### Grey goo

*Grey goo* refers to a hypothetical end-of-the-world event involving molecular nanotechnology in which out-of-control self-replicating robots consume all living matter on Earth while building more of themselves (a scenario known as ecophagy).

The term is usually used in a science fiction context. In a worst-case scenario, all of the matter in the universe could be turned into goo (with "goo" meaning a large mass of replicating nanomachines lacking large-scale structure, which may or may not actually appear goo-like), killing the universe's inhabitants. The disaster is posited to result from an accidental mutation in a self-replicating nanomachine used for other purposes, or possibly from a deliberate doomsday device.

It is thus worth noting that grey goo need not be grey or gooey. They could be like, for all purposes, a plant or bacteria. It is only the result of their ecophagy that would resemble grey goo. One convenient analogy for the grey goo problem is to consider bacteria as the most perfect example of biological nanotechnology. As they have not reduced the world to living goo, some consider it unlikely that some artificial construct will manage to do so with grey goo.

Even so, some people argue that living goo, or even a combination of nanotechnology and biotechnology to create organic replicators, is a more realistic threat than grey goo. Arguing that bacteria are ubiquitous and extraordinarily powerful, Bill Bryson (2003) says that the Earth is "their planet" and that we're only allowed to exist on it because "they allow us to". Margulis and Sagan (1995) go further, arguing that all organisms, having descended from bacteria, *are* in a sense bacteria. Many kinds of bacteria are in fact essential for human life and are found in large quantities in the human digestive tract, in a symbiotic relationship.

Thus a living goo could be a multicellular organism that obtains its raw materials to grow through ecophagy, and then grows through a process of exponential assembly such as cell division.

It is unclear whether the molecular nanotechnology would be capable of creating grey goo at all. Among other common refutations, theorists suggest that the very size of nanoparticles inhibits them from moving very quickly. While the biological matter that composes life releases significant amounts of energy when oxidised, and other sources of energy such as sunlight are available, this energy might not be sufficient for the putative nanorobots to outcompete existing organic life that already uses those resources, especially considering how much energy nanorobots would use for locomotion. If the nanomachine was itself composed of organic molecules, then it might even find itself being preyed upon by preexisting bacteria and other natural life forms.

If nanorobots were built of inorganic compounds or made much use of elements that are not generally found in living matter, then they would need to use much of their metabolic output for fighting entropy as they purified (reduce sand to silicon, for instance) and synthesized the necessary building blocks. There would be little chemical energy available from inorganic matter such as rocks because, aside from a few exceptions (coal, for example) it is mostly well-oxidized and sitting in a free-energy minimum.

Assuming a molecular nanotechnological replicator were capable of causing a grey goo disaster, safety precautions might include programming them to stop reproducing after a certain number of generations, designing them to require a rare material that would be sprayed on the construction site before their release, or requiring constant direct control from an external computer. Another possibility is to encrypt the memory of the replicators in such a way that any changed copy is overwhelmingly likely to decrypt to non-functioning static.

In Britain, the Prince of Wales called upon the Royal Society to investigate the "enormous environmental and social risks" of nanotechnology in a planned report, leading to much delighted media commentary on grey goo. The Royal Society's report on nanoscience was released on 29 July 2004.

Recently, new analysis has shown that the danger of grey goo is far less likely than originally thought. However, other long-term major risks to society and the environment from nanotechnology have been identified. Drexler has made a somewhat public effort to retract his grey goo hypothesis, in an effort to focus the debate on more realistic threats associated with knowledge-enabled nanoterrorism and other misuses.

## Ground State

The lowest-energy state of a system. The electronic ground state of a system cannot reduce its energy by an electronic transition, but may contain vibrational energy (kinetic and potential energy associated with the motions and positions of its atoms); extended systems at ordinary temperatures are always vibrationally excited, and so "ground state" is often taken to mean "electronic ground state."

## Ground

To make electrical connection to the earth or to the chassis of a device (verb); the connection point so used (noun).

## Group Velocity

In wave propagation, the speed of the waveform (e.g., of a peak) can be different from the speed of a group of waves (e.g., of a set of ripples in water). The latter is the *group velocity*, and is the speed of propagation of information and wave energy. The waveform speed is the *phase velocity*.

## Group

A set of linked atoms in a molecule; a defined substructure. Typically, a set that is usefully regarded as a unit in chemical reactions of interest.

## GUI

Graphical User Interface — hardware, software, and firmware that produces the display on modern personal computers.

## Guy Fawkes Scenario

If nanotechnology becomes widely available, it might become trivial for anyone to committ acts of terrorism (such as making nanomachines build a large amount of explosives under government buildings a la Guy Fawkes). This would either force strict control over nanotechnology (hard) or a decentralized mode of organization.

## Gyroscope

A gyroscope is a device for measuring or maintaining orientation, based on the principle of conservation of angular momentum. In physics this is also known as gyroscopic inertia or rigidity in space. The essence of the device is a spinning wheel on an axle. The device, once spinning, tends to resist changes to its orientation due to the angular momentum of the wheel.

Within mechanical combinations or devices constituting portions of machines, a conventional *gyroscope* is a mechanism comprising a rotor journaled to spin about one axis, the journals of the rotor being mounted in an inner gimbal or ring, the inner gimbal being journaled for oscillation in an outer gimbal which in turn is journaled for oscillation relative to a support. The outer gimbal or ring is mounted so as to pivot about an axis in its own plane determined by the support. The outer gimbal possesses one degree of rotational freedom and its axis possesses none. The inner gimbal is mounted in the outer gimbal so as to pivot about an axis in its own plane which axis is always normal to the pivotal axis of the outer gimbal. Simple gyro wheel. Reaction arrows about the output axis correspond to forces applied about the input axis. (e.g., CW input results in CCW output)

The axle of the spinning wheel defines the spin axis. The inner gimbal possesses two degrees of rotational freedom and its axis possesses one. The rotor is journaled to spin about an axis which is always normal to the axis of the inner gimbal. Hence the rotor possesses three degrees of rotational freedom and its axis possesses two. The wheel responds to a force applied about the input axis by a reaction force about the output axis. The 3 axes are perpendicular, and this cross-axis response is the simple essence of the gyroscopic effect.

A gyroscope flywheel will roll or resist about the output axis depending upon whether the output gimbals are of a free- or fixed- configuration. Examples of some free-output-gimbal devices would be the attitude reference gyroscopes used to sense or measure the pitch, roll and yaw attitude angles in a spacecraft or airplane, and the front wheel of a motorcycle. Countersteering is how motorcycles turn corners using the gyroscopic roll reaction of the spinning front wheel.

The center of gravity of the rotor can be in a fixed position. The rotor simultaneously spins about one axis and is capable of oscillating about the two other axes, and thus, except for its inherent resistance due to rotor spin, it is free to turn in any direction about the fixed point. Some gyroscopes have mechanical equivalents substituted for one or more of the elements, e.g., the spinning rotor may be suspended in a fluid, instead of being pivotally mounted in gimbals. A control moment gyroscope (CMG) is an example of a fixed-output-gimbal device that is used on spacecraft to hold or maintain a desired attitude angle or pointing direction using the gyroscopic resistance force.

In some special cases, the outer gimbal (or its equivalent) may be omitted so that the rotor has only two degrees of freedom. In other cases, the center of gravity of the rotor may be offset from the axis of oscillation, and thus the center of gravity of the rotor and the center of suspension of the rotor may not coincide.

The gyroscope effect was discovered in 1817 by Johann Bohnenberger and invented and named in 1852 by Léon Foucault for an experiment involving the rotation of the Earth. Foucault's experiment to see (*skopeein*, to see) the Earth's rotation (*gyros*, circle or rotation) was unsuccessful due to friction, which effectively limited each trial to 8 to 10 minutes, too short a time to observe significant movement. In the 1860s, however, electric motors made the concept feasible, leading to the first prototype gyrocompasses; the first functional marine gyrocompass was developed between 1905 and 1908 by German inventor Hermann Anschütz-Kaempfe. The American Elmer Sperry followed with his own design in 1910, and other nations soon realized the military importance of the invention—in an age in which naval might was the most significant measure of military power—and created their own gyroscope industries. The Sperry Gyroscope Company quickly expanded to provide aircraft and naval stabilizers as well, and other gyroscope developers followed suit.

In the first several decades of the 20th century, other inventors attempted (unsuccessfully) to use gyroscopes as the basis for early black box navigational systems by creating a stable platform from which accurate acceleration measurements could be performed (in order to bypass the need for star sightings to calculate position). Similar principles were later employed in the development of inertial guidance systems for ballistic missiles.

A gyroscope exhibits a number of behaviours including precession and nutation. Gyroscopes can be used to construct gyrocompasses which complement or replace magnetic compasses (in ships, aircraft and spacecraft, vehicles in general), to assist in stability (bicycle, Hubble Space Telescope, ships, vehicles in general) or be used as part of an Inertial guidance system. Gyroscopic effects are used in toys like yo-yos and Powerballs. Many other rotating devices, such as flywheels, behave gyroscopically although the gyroscopic effect is not used.

It follows from this that a torque Ä applied perpendicular to the axis of rotation, and therefore perpendicular to L, results in a motion perpendicular to both Ä and L. This motion is called *precession*.

Precession can be demonstrated by placing a spinning gyroscope with its axis horizontal and supported loosely at one end. Instead of falling, as might be expected, the gyroscope appears to defy gravity by remaining with its axis horizontal, even though one end of the axis is unsupported. The free end of the axis slowly describes a circle in a horizontal plane. This effect is explained by the above equations. The torque on the gyroscope is supplied by a couple of forces: gravity acting downwards on the device's centre of mass, and an equal force acting upwards to support one end of the device. The motion resulting from this torque is not downwards, as might be intuitively expected, causing the device to fall, but perpendicular to both the gravitational torque (downwards) and the axis of rotation (outwards from the point of support), i.e. in a forward horizontal direction, causing the device to rotate slowly about the supporting point.

As the second equation shows, under a constant torque due to gravity, the gyroscope's speed of precession is inversely proportional to its angular momentum. This means that, as friction causes the gyroscope's spin to slow down, the rate of precession increases. This continues until the device is unable to rotate fast enough to support its own weight, when it stops precessing and falls off its support.

By convention, these three vectors, torque, spin, and precession, are all oriented with respect to each other according to the right-hand rule. The right-hand rule is a handy trick for keeping track of vector orientation. In this case it works in two ways. First the direction of these vectors is determined by the right hand rule: fingers of the right hand wrapping in direction of rotation leave the thumb of the right hand pointing in direction of the corresponding vector. Then the fingers of the right hand initially oriented in the direction of the spin vector and bent in the direction of the moment vector leave the thumb pointing in the direction of the precession vector.

# H

## Harmonic Oscillator

A system in which a mass is subject to a linear restoring force, like an ideal spring. A harmonic oscillator vibrates at a fixed frequency, independent of amplitude.

## Heat

Heat, symbolized by $Q$, is defined as *energy in transit*. Generally, heat is a form of energy associated with the motion of atoms, molecules and other particles which comprise matter. Heat can be created by chemical reactions (such as burning), nuclear reactions (such as fusion taking place inside the Sun), electromagnetic dissipation (as in electric stoves), or mechanical dissipation (such as friction). Heat can be transferred between objects by radiation, conduction and convection. Temperature, defined as the measure of an object to spontaneously give up energy, is used to indicate the level of elementary motion associated with heat. Heat can only be transferred between objects, or areas within an object, with different temperatures, and then only in the direction of the colder body.

The first to have put forward a semblance of a theory on heat was the Greek philosopher Heraclitus who lived around 500 BC in the city of Ephesus in Ionia, Asia Minor. He became famous as the "flux and fire" philosopher for his proverbial utterance: "All things are flowing." Heraclitus argued that the three principle elements in nature were fire, earth, and water. Of these three, however, fire is assigned as the central element controlling and modifying the other two. The universe was postulated to be in a continuous state of state of flux or permanent condition of change as a result of transformations of fire. Heraclitus summarized his philosophy as: "All things are an exchange for die"

### Heisenberg Uncertainty Principle

A quantum-mechanical principle with the consequence that the position and momentum of an object cannot be precisely determined. The Heisenberg principle helps determine the size of electron clouds, and hence the size of atoms.

### HeLa Cell

Line of human epithelial cells that grows vigorously in culture. Derived from a human cervical carcinoma.

### Heme

Cyclic organic molecule containing an iron atom that carries oxygen in hemoglobin and carries an electron in cytochromes.

### Hemidesmosome

Specialized cell junction between an epithelial cell and the underlying basal lamina.

### Hemoglobin

The major protein in red blood cells that associates with $O_2$ in the lungs by means of a bound heme group.

### Henry

Unit of inductance. One henry (H) is the inductance of a closed circuit in which an electromotive force of 1 volt is produced when the electric current in the circuit varies uniformly at the rate of 1 ampere per second.

### Hermes

The first satellite to experiment with small satellite dishes for television. This use of small dishes made live news reports from remote locations possible. The Canadian Hermes satellite was launched on January 17, 1976.

### Heterochromatin

Region of a chromosome that remains unusually condensed and transcriptionally inactive during interphase.

### Heterodimer

Protein complex composed of two different polypeptide chains.

## Heuristics

Rules of thumb used to guide one in the direction of probable solutions to a problem.

## High Surface Area Materials

Materials which have large surface to volume ratio. Consider for example a chunk of material in the shape of a sphere of radius R. Its surface to volume ratio is 3/R.

Thus, for smaller pieces of materials, the surface to volume ratio is larger. For pieces of matter on a nanometer scale (nanoparticles), the contribution of the surface to the overall properties of such matter becomes very important. For example, in a nanoscopic particle with a diameter of 5 nm, about half of its atoms are on its surface (low-coordination-number atoms). This is why the techniques and methods of surface science are needed to describe properties of nanoscopic matter. Even the macroscopic pieces of nanostructured materials can have high surface area, since their nanoscopic constituents can arrange in such a way to still have a large exposed surface.

For example, single-walled carbon nanotube materials consist of carbon nanotube bundles. The bundles are made of several tens of carbon nanotubes and thus have a large exposed surface. Bundles are in a macroscopic material organized in such a way that their large surface still remains exposed (a particular bundle is not shielded by other bundles in the material). High surface area materials are of interest for catalysis, due to the fact that catalysis proceeds on the surface of materials.

Porous materials such as zeolites also have a large area of exposed surface. Such materials can also be used for separation of different gases (purification of gases, removal of ultra-fine contaminants from air and water) since, due to geometrical constraints, molecules of a particular gas can enter the nanoscopic pores, while those of another gas cannot due to the fact that they are to big to fit in.

## Histonatation

In medical nanorobotics, locomotion (swimming) through tissues by a nanorobot.

## Histone

A protein present in the eukaryotic nucleus that is bound to the DNA at regularly spaced intervals.

## Homogeneous and Heterogeneous Assays

A homogeneous assay does not require a separation step to remove free antigen from bound antigen and relies upon the fact that the function of the label is modified upon binding, leading to a change in signal intensity. Because of high background signal a heterogeneous approach incorporating a separation step of bound and unbound makes the detection limit lower, approaching the values obtained by RIA. The homogeneous assay is less technically demanding.

## Hubble Space Telescope

A satellite launched by NASA in April, 1990. It is the largest telescope ever built to go into space. It is known as an astronomical observatory—in both senses of the word! Hubble is extremely powerful, able to look deep into space to study distant galaxies and stars.

## Human Genome Project

A research initiative that has mapped the entire human genome.

## Human Tissue Engineering

No universally agreed definition exists. But a Commission working hypothesis suggests a human tissue engineered product means any autologous (emanating from the patient himself) or allogeneic (coming from another human being) product which: contains, consists of, or results in engineered human cells or tissues; and has properties for, or is presented as having properties for, the regeneration, repair or replacement of tissue, where the new tissue or cells, in whole or in part, are structurally and functionally analogous to the original tissue that is being regenerated, repaired or replaced.

## Hydrogen Bond

A bond formed between hydrogen and two other atoms. A hydrogen atom covalently bound to an electronegative atom (e.g., nitrogen, oxygen) has a significant positive charge and can form a weak bond to another electronegative atom; this is termed a hydrogen bond.

### Hydrophobic Force

Water molecules are linked by a network of hydrogen bonds. A nonpolar, nonwetting, surface (e.g., wax) cannot form hydrogen bonds. To form their full complement of hydrogen bonds, the nearby water molecules must form a more orderly (hence lower entropy) network. This both increases free energy and causes forces that tend to draw hydrophobic surfaces together across distances of several nanometers.

### Hydrophobic

Literally—water fearing, from the Greek *hydro*—"water" and *phobo*—"fear". The hydrophobic effect is the entropy driven force that causes oil to separate from water. It is notoriously strong, though not as strong as covalent forces. This force is one of the main determinants of the structure of globular protein molecues, since the hydrophilic (water loving) parts of the molecule tend to surround the hydrophobic parts that cluster in the center, away from the aqueous (polar) solvent.

### Hyperpigmentation

Hyperpigmentation is a common condition in which some patches of skin turn darker in color. This is a harmless condition caused when there is too much brown pigment, called melanin, in the skin. This condition can affect people in all races. Age spots, sometimes called liver spots, are a form of hyperpigmentation. They usually occur because of damage to the skin from the sun. Doctors call these spots solar lentigines. The small, dark spots are found generally on the hands and face, but any area exposed to a lot of sun can be affected. There are two types of spots that are similar to age spots, but they cover larger areas of skin. These are referred to as melasma or chloasma spots and, while they are similar to age spots, they are a result of hormonal changes. Some pregnant women overproduce melanin and they get a condition called "mask of pregnancy" on their faces or abdomens. Women who take birth control pills may also develop hyperpigmentation since their bodies react, hormonally, as if they were pregnant. Hyperpigmentation is not the only cause of skin color change. Acne can cause darkening of the skin and so can skin injuries and some surgery. Freckles are darkened places on the skin of the face and arms and is a hereditary condition. Any darkened skin patch can get even darker when that area of the skin is exposed to the sun because melanin absorbs ultraviolet rays from the sun in order to protect the skin from overexposure. This is actually what we refer to as "tanning."

There are a variety of prescription creams that can help to lighten the darkened patches of skin. They do this by slowing melanin production so that the patches fade. Laser treatments are also effective at removing hyperpigmentation and often can remove the pigmented areas without leaving any scars.

## Hysteresis

Hysteresis is a property of systems (usually physical systems) that do not instantly follow the forces applied to them, but react slowly, or do not return completely to their original state: that is, systems whose states depend on their immediate history. For instance, if you push on a piece of putty it will assume a new shape, and when you remove your hand it will not return to its original shape, or at least not immediately and not entirely. The term derives from an ancient Greek word, meaning 'deficiency'. The phenomenon was identified, and the term coined, by Sir James Alfred Ewing in 1890.

Hysteresis phenomena occur in magnetic and ferromagnetic materials, as well as in the elastic and electromagnetic behavior of materials, in which a lag occurs between the application and the removal of a force or field and its subsequent effect. Electric hysteresis occurs when applying a varying electric field, and elastic hysteresis occurs in response to a varying force. The term "hysteresis" is sometimes used in other fields, such as economics or biology. In such cases it describes a memory or lagging effect in which the order of previous events can influence the order of subsequent events.

The word "lag" above should not necessarily be interpreted as a time lag. After all, even relatively simple linear systems such as an electric circuit containing resistors and capacitors exhibit a time lag between the input and the output. For most hysteretic systems, there is a very short time scale when its dynamic behavior and various related time dependences are observed. In magnetism, for example, the dynamic processes occurring on this very short time scale have been referred to as Barkhausen jumps. If observations are carried out over very long periods of time, creep or slow relaxation typically toward true thermodynamic equilibrium (or other types of equilibria that depend on the nature of the system) can be noticed. When observations are carried out without regard for very fast dynamic phenomena or very slow relaxation phenomena, the system appears to display irreversible behavior whose rate is practically independent of the driving force rate. This rate-independent irreversible behavior is the key feature that distinguishes hysteresis from most other dynamic processes in many systems. If the displacement of a system with hysteresis is plotted on a graph against the applied force, the resulting curve is in the form of a loop. In contrast, the curve for a system without hysteresis is a single, not necessarily straight, line. Although the hysteresis loop depends on the material's physical properties, there is no complete theoretical description that explains the phenomenon. The family of hysteresis loops, from the results of different applied varying voltages or forces, form a closed space in three dimensions, called the hysteroid.

Hysteresis was initially seen as problematic, but is now thought to be of great importance in technology. For instance, the properties of hysteresis are applied when constructing permanent memory for computers: hysteresis allows most superconductors to operate at the high currents needed to create strong magnetic fields. Hysteresis is also important in living systems. Many critical processes occurring in living (or dying) cells use hysteresis to help stabilize them against the various effects of random chemical fluctuations.

Some early work on describing hysteresis in mechanical systems was performed by James Clerk Maxwell. Subsequently, hysteresis models have received significant attention in the works of Preisach, Neel and Everett in connection with magnetism and absorption. More formal mathematical theory of systems with hysteresis was developed in 1970s, by a group of Russian mathematicians, which was led by Mark Krasnosel'skii, one of the founders of nonlinear analysis. He suggested an investigation of hysteresis phenomena using the theory of nonlinear operators.

*Informal Definition*

The phenomenon of hysteresis can conceptually be explained as follows. A system can be divided into subsystems or *domains*, much larger than an atomic volume but still microscopic. Such domains normally occur in ferroelectric and ferromagnetic systems, since individual dipoles tend to group with each other, forming a small isotropic region. Each of the system's domains can be shown to have a metastable state. The metastable domains can in turn have two or more substates. Such a *metastable state* fluctuates widely from domain to domain, but the average represents the configuration of lowest energy. The *hysteresis* is simply the sum of all domains, or the sum of all metastable states.

Hysteresis is well known in ferromagnetic materials. When an external magnetic field is applied to a ferromagnet, the ferromagnet absorbs some of the external field. Even when the external field is removed, the magnet will retain some field: it has become *magnetized*.

A family of B-H loops for grain-oriented electrical steel ($B_R$ denotes *remanence* and $H_C'$ is the *coercivity*.

The relationship between magnetic field strength (H) and magnetic flux density (B) is not linear in such materials. If the relationship between the two is plotted for increasing levels of field strength, it will follow a curve up to a point where further increases in magnetic field strength will result in no further change in flux density.

This condition is called magnetic saturation.

If the magnetic field is now reversed and increased linearily, the plotted relationship will again follow a similar curve back towards and beyond zero flux density but offset from the original curve by an amount called the *remanent flux density* or *remanence.*

If this relationship is plotted for all strengths of applied magnetic field the result is a sort of *S- shaped* loop. The 'thickness' of the middle bit of the S describes the amount of hysteresis, related to the coercivity of the material.

Its practical effects might be, for example, to cause a relay to be slow to release due to the remaining magnetic field continuing to attract the armature when the applied electric current to the operating coil is removed.

This is also a very important effect in magnetic tape and other magnetic storage media like hard disks. In these materials it would seem obvious to have one polarity represent a bit, say north for 1 and south for 0. However, if you want to change the storage from one to the other, the hysteresis effect requires you to know what was already there, because the needed field will be different in each case. In order to avoid this problem, recording systems first overdrive the entire system into a known state using a process known as bias. Analogue magnetic recording also uses this technique. Different materials require different biasing, which is why there is a selector for this on the front of most cassette recorders.

In order to minimize this effect and the power losses associated with it, ferromagnetic substances with low coercivity and low hysteresis loss are used, like permalloy.

*Electrical hysteresis*

Electrical hysteresis typically occurs in ferroelectric material, where domains of polarization contribute to the total polarization. Polarization is the electrical dipole moment.

*Liquid-solid phase transitions*

Hysteresis manifests itself in state transitions when melting temperature and freezing temperature do not agree. For example, agar melts at 85 °C and solidifies from 32 to 40 °C. This is to say that once agar is melted at 85 degrees, it retains a liquid state until cooled to 40 degrees Celsius. Therefore, from the temperatures of 40 to 85 degrees Celsius, agar can be either solid or liquid, depending on which state it was before.

*Matric potential hysteresis*

The relationship between matric water potential and water content is the basis of the water retention curve. Matric potential measurements (¨$_m$) are converted to volumetric water content (¸) measurements based on a site or soil specific calibration curve. Hysteresis is a source of water content measurement error. Matric potential hysteresis arises from differences in wetting behaviour causing dry medium to re-wet; that is, it depends on the saturation history of the porous medium. Hysteretic behaviour means that, for example, at a matric potential (¨$_m$) of 5kPa, the volumetric water content of a fine sandy soil matrix could be anything between 8% to 25%.

Tensiometers are directly influenced by this type of hysteresis. Two other types of sensors used to measure soil water matric potential are also influenced by hysteresis effects within the sensor itself. Resistance blocks, both nylon and gypsum based, measure matric potential as a function of electrical resistance. The relation between the sensor's electrical resistance and sensor matric potential is hysteretic. Thermocouples measure matric potential as a function of heat dissipation. Hysteresis occurs because measured heat dissipation depends on sensor water content, and the sensor water content–matric potential relationship is hysteretic. As of 2002, only desorption curves are usually measured during calibration of soil moisture sensors. Despite the fact that it can be a source of significant error, the sensor specific effect of hysteresis is generally ignored.

*Energy*

When hysteresis occurs with extensive and intensive variables, the work done on the system is the area under the hysteresis graph.

*User interface design*

The field of user interface design has borrowed the term hysteresis to refer to times when the state of the user interface intentionally lags behind the apparent user input. For example, a menu that was drawn in response to a mouse-over event may remain on-screen for a brief moment after the mouse has moved out of the trigger region and the menu region. This allows the user to move the mouse directly to an item on the menu, even if part of that direct mouse path is outside of both the trigger region and the menu region.

*Electronics*

Hysteresis can be used to filter a signal so that the output reacts slowly by taking recent history into account. For example, a thermostat controlling a heater may turn the heater on when the temperature drops below A degrees, but not turn it off until the temperature rises above B degrees. Thus the on/off output of the thermostat to the heater when the temperature is between A and B depends on the history of the temperature. This prevents rapid switching on and off as the temperature drifts around the set point.

A Schmitt trigger is a simple electronic circuit that also exhibits this property. Often, some amount of hysteresis is intentionally added to an electronic circuit (or digital algorithm) to prevent unwanted rapid switching.

*Neuroscience*

The property by which some neurons do not return to their basal conditions from a stimulated condition immediately after removal of the stimulus is an example of hysteresis.

# I

## Immune System

The human immune system is a truly amazing constellation of responses to attacks from outside the body. It has many facets, a number of which can change to optimize the response to these unwanted intrusions. The system is remarkably effective, most of the time. This note will give you a brief outline of some of the processes involved.

An *antigen* is any substance that elicits an immune response, from a virus to a sliver. The immune system has a series of dual natures, the most important of which is self/non-self recognition. The others are: general/specific, natural/adaptive = innate/acquired, cell-mediated/humoral, active/passive, primary/secondary. Parts of the immune system are antigen-specific (they recognize and act against particular antigens), systemic (not confined to the initial infection site, but work throughout the body), and have memory (recognize and mount an even stronger attack to the same antigen the next time).

Self/non-self recognition is achieved by having every cell display a marker based on the major histocompatibility complex (MHC). Any cell not displaying this marker is treated as non-self and attacked. The process is so effective that undigested proteins are treated as antigens. Sometimes the process breaks down and the immune system attacks self-cells. This is the case of *autoimmune diseases* like multiple sclerosis, systemic lupus erythematosus, and some forms of arthritis and diabetes. There are cases where the immune response to innocuous substances is inappropriate. This is the case of allergies and the simple substance that elicits the response is called an *allergen*.

### Immuno-electron Microscopy

This technique allows the investigator to identify antibody/ antigen complexes that localize to a particular subcellular organelle or compartment by using the Protein A gold technique. The tissue is first fixed in 4.0% para-formaldehyde and processed for embedding in a low temperature embedding resin (Lowicryl).

Paraffin embedded tissue (same tissue) sections are cut and screened by light microscopy using a streptavidin-biotin technique to determine the appropriate antibody concentration to be used for the EM immunogold procedure.

Ultrathin sections from the same block(s) are cut, incubated with primary antibody and then later incubated with Protein A gold particles (size range is 5 nm to 20 nm). The gold particles bind to the Fc portion of the antibody and are detected by EM. A variation of the large block "pop-off" technique for immunoelectron microscopy is also available.

### Immunoglobulin

An antibody molecule. Higher vertebrates have five classes of immunoglobulin—IgA, IgD, IgE, IgG, and IgM—each with a different role in the immune response.: : immunoglobulin like (Ig-like) domain: : Characteristic protein domain of about 100 amino acids that is found in antibody molecules and in many other proteins that form the Ig superfamily.

### Immunohistochemistry

Histochemical localization of immunoreactive substances using labeled antibodies as reagents. Immunohistochemistry involves using antibodies (typically visualized via an enzyme-linked antibody assay) that specifically bind to proteins of interest. This method allows one not only to assess levels of a protein but also to localize the protein within cells in the tissue sample.

### Impedance

The complex ratio of a force-like quantity (force, pressure, voltage, temperature, or electric field) to a corresponding related velocity-like quantity (velocity, volume velocity, current, heat flow, or magnetic field strength).

## Inductance

The property of an electric circuit which tends to oppose change in current in the circuit. One henry (H) is the inductance of a closed circuit in which an electromotive force of 1 volt is produced when the electric current in the circuit varies uniformly at the rate of 1 ampere per second.

## Inductor

An inductor is a passive electrical device employed in electrical circuits for its property of inductance. An inductor can take many forms.

Inductance (measured in henries) is an effect which results from the magnetic field that forms around a current carrying conductor. Electrical current through the conductor creates a magnetic flux proportional to the current. A change in this current creates a change in magnetic flux that, in turn, generates an electromotive force (emf) that acts to oppose this change in current. Inductance is a measure of the generated emf for a unit change in current. For example, an inductor with an inductance of 1 henry produces an emf of 1 V when the current through the inductor changes at the rate of 1 ampere per second. The inductance of a conductor is increased by coiling the conductor such that the magnetic flux encloses (links) all of the coils (turns). Additionally, the magnetic flux linking these turns can be increased by coiling the conductor around a material with a high permeability.

Electrical current can be modeled by fluid flow, much like water through pipes where the rate of flow corresponds to electric current and pressure corresponds to potential difference. Hence, the inductor can be modeled by the flywheel effect of a turbine rotated by the flow. As can be demonstrated intuitively and mathematically, this mimics the behavior of an electrical inductor; current is the integral of voltage and in cases of a sudden interruption of flow the turbine will generate a high pressure across the blockage, etc. Magnetic interactions such as transformers can be modelled by two or more separate turbines on a common shaft. For example, a 2:1 transformer can be modelled as two coupled turbines. For a given rotational speed the larger will produce a flow of x litres per second at a pressure drop of y pascals and the smaller turbine will produce a flow of x/2 litres per second at 2y pascals. In the turbine model, the angular velocity of the rotating parts corresponds to magnetic flux in the core of an inductor, and the torque corresponds to the rate of change of flux.

An inductor is usually constructed as a coil of conducting material, typically copper wire, wrapped around a core either of air or of ferromagnetic material. Core materials with a higher permeability than air confine the magnetic field closely to the inductor, thereby increasing the inductance. Inductors come in many shapes. Most are constructed as enamel coated wire wrapped around a ferrite bobbin with wire exposed on the outside, while some enclose the wire completely in ferrite and are called "shielded". Some inductors have an adjustable core, which enables changing of the inductance. Small inductors can be etched directly onto a printed circuit board by laying out the trace in a spiral pattern. Small value inductors can also be built on integrated circuits using the same processes that are used to make transistors. In these cases, aluminum interconnect is typically used as the conducting material. However, practical constraints make it far more common to use a circuit called a "gyrator" which uses a capacitor and active components to behave similarly to an inductor. Inductors used to block very high frequencies are sometimes made with a wire passing through a ferrite cylinder or bead.

## Inflammatory Response

When a bacterial infection is established in the body, it is the function of the immune system to control or eradicate it. The initial reaction of the immune system to an infection differs, depending on the site which has been invaded, and on the nature of the invader.

There can be many "triggers", that can kick the immune system into action. Here are some of the ways in which the immune system can be activated.

## Infrared Spectromicroscopy

Synchrotron Radiation is a powerful tool in research, broadly applied in the VUV and x-ray spectral regions. The three main advantages of synchrotron radiation are: broadband characteristics, high collimation and the possibility to calculate its properties with Schwinger's equation. Synchrotron radiation can also be superior in the infrared spectral region.

## Infrared

Infrared (IR) radiation is electromagnetic radiation of a wavelength longer than that of visible light, but shorter than that of radio waves. The name means "below red" (from the Latin *infra*, "below"), red being the color of visible light of longest wavelength. Infrared radiation spans three orders of magnitude and has wavelengths between approximately 750 nm and 1 mm.

However, these terms are not precise, and are used differently in various studies i.e. near (0.75–5 µm) / mid (5–30 µm) / long (30–1,000 µm). Especially at the telecom wavelengths the spectrum is further subdivided into individual bands, due to limitations of detectors, amplifiers and sources.

Infrared radiation is popularly known as "heat" or perhaps "heat radiation," since many physics teachers traditionally attribute all radiant heating to infrared light. This is wrong, and is a very widespread misconception. Light or electromagnetic waves of any frequency will heat surfaces which absorb it. IR light from the sun only accounts for 50% of the heating of the Earth, the rest is caused by visible light.

Green lasers can char paper, incandescently hot objects put out visible radiation, and ice cubes emit mostly microwaves. However, it is true that objects at room temperature will emit radiation mostly concentrated in the mid-infrared band

The common nomenclature is justified by the different human response to this radiation: near infrared is the region closest in wavelength to the radiation detectable by the human eye, mid and far infrared are progressively further from the visible regime. Other definitions follow different physical mechanisms (emission peaks, vs. bands, water absorption) and the newest follow technical reasons (The common silicon detectors are sensitive to about 1,050 nm, while InGaAs sensitivity starts around 950 nm and ends between 1,700 and 2,600 nm, depending on the specific configuration). Unfortunately the international standards for these specifications are not currently available.

The boundary between visible and infrared light is not precisely defined. The human eye is markedly less sensitive to light above 700 nm wavelength, so longer frequencies make insignificant contributions to scenes illuminated by common light sources. But particularly intense light (e.g., from lasers) can be detected up to approximately 780 nm, and will be perceived as red light. The onset of infrared is defined (according to different standards) at various values typically between 700 nm and 780 nm.

### Inhibitor

A chemical substance that, when added in relatively low concentrations, slows down a chemical reaction.

### Inline Universities

(As opposed to online universities), nanocomputer implants serving to increase intelligence and education of their owners, essentially turning them into walking universities.

### Inmessaging

In medical nanorobotics, conveyance of information from a source external to the human body, or external to working nanodevices, to a receiver located inside the human body.

### Input Device

A digitizer, scanner, line art, clip art, keyboard, video camera or any other device that is used to generate and send a design or other instruction into a computer for eventual production on an output device.

## Insertion Point (in lithography context)

Adaptation of a new lithography technique is referred to as the insertion point of that technique.

## Insulator

An electrical insulator resists the flow of electricity. Application of a voltage difference across a good insulator results in negligible electrical current. In comparison, a conductor allows current to flow readily. Controlling the flow of current in electrical wiring and electronic circuits requires both insulators and conductors. For example, wires typically consist of a current-carrying metallic core sheathed in an insulating coating. Resistivity is the measure of a material's effectiveness in resisting current flow. Materials with resistivities higher than $10^8$ ohm-m are usually considered to be good insulators; these include glass, rubber, and many plastics. Resistivities as high as $10^{16}$ ohm-m can be achieved in exceptional insulating materials. Normal conductors may have resistivities as low as $10^{-8}$ ohm-m. The enormous variation in room-temperature resistivity is one of the largest for any physical attribute of matter.

The charge carriers responsible for current in most conductors are electrons, moving relatively freely in a metal. In insulators, electrons cannot move freely. When atoms of simple metals combine to form a solid, the outer valence electrons become free for conduction. In an ideal insulator, all electrons stay tightly bound to the atoms, so there are no electrons that can be readily moved through the material for conduction. A more complete understanding of insulators and conductors requires consideration of electronic band structure. The electrons in an isolated atom possess discrete energies, a consequence of quantum mechanics. These discrete levels evolve into bands of allowed energies when the atoms condense into a solid. Forbidden regions separate the allowed bands, as shown schematically in the figure. The electrons in the solid fill in the bands, from lower to higher energy.

The distinction between an

insulator and a conductor lies in how the electrons fill in the allowed bands. For a simple metal, the highest band containing electrons will be only half full. The thermal energy (at ordinary temperatures) will be sufficient to generate conduction electrons — electron states of slightly higher energy are available in the incompletely filled band.

In comparison, the highest energy band containing electrons is completely full in a good insulator. The thermal energy of the electrons is not sufficient for promotion from this band, known as the valence band, to the next band with available energy states, known as the conduction band. The gap between the valence and conduction band, known as the band gap, is at least several electron-volts (eV) wide in an insulator — thermal electron energies are 100 times smaller.

The distinction between an insulator and a semiconductor is one of degree. Although both have completely filled valence bands at 0 K, the band gap of a semiconductor is smaller than an insulator. For narrower band gaps, thermal energy is more capable of promoting electrons into the conduction band.

Insulators play a critical role in many aspects of technology, from large scale to the microscopic. Electrical power transfer relies on high voltage transmssion lines for which insulators are required to prevent losses to ground. High voltage transformers rely on special insulating oils. Dielectric materials enhance the charge storage of a capacitor. Even bits of information in a computer memory require thin insulators in the form of oxide layers in increasingly smaller transistor circuits

## Intelligent Agent

In computer science, an intelligent agent (IA) is a software agent that exhibits some form of artificial intelligence that assists the user and will act on their behalf, in performing repetitive computer-related tasks. While the working of software agents used for *operator assistance* or data mining is often based on fixed pre-programmed rules, "intelligent" here implies the ability to adapt and learn.

In some literature IAs are also referred to as *autonomous intelligent agents*, which means they act independently, and will learn and adapt to changing circumstances. According to Nicola Kasabov IA systems should exhibit the following characteristics: (i) learn and improve through interaction with the environment (embodiment); (ii) adapt online and in real time; (iii) learn quickly from large amounts of data; (iv) accommodate new problem solving rules incrementally; (v) have memory based exemplar storage and retrieval capacities; (vi) have parameters to represent short and long term memory, age, forgetting, etc. and (vii) be able to analyze itself in terms of behavior, error and success

### Interdisciplinarity

A view or approach which crosses the borders of traditional disciplines (e.g. physics, chemistry, biology, art ...). It is often said that nanotechnology requires an interdisciplinary approach. This is partly due to the fact that the nanoscience itself is a term representing various sciences dealing with very different object which have in common their nanometric size. Some say that a 'new breed' of researches, educated in a range of traditional disciplines, is needed to make real progress in nanotechnology.

### Interface

In physical terms, an interface is the boundry between two phases, for instance between a solid and liquid or between a liquid and gas.

### Intermolecular

Between more than one molecule. Describes an interaction (e.g., a chemical reaction) between different molecules.

### Internal Energy

The sum of the kinetic and potential energies (including electromagnetic field energies) of the particles that make up a system.

### Interstitial Diffusion

A diffusion mechanism that causes atomic motion from interstitial site to interstitial site.

### Intrinsic

Characterizes pure undoped semiconductor; electrical conductivity depends only on temperature and the band gap energy.

### Ion Exchange

A chemical reaction involving the exchange of hydrated ions in a solid for similarly charged but different element ions in solution.

### Ion Microscopy

Use of the Secondary Ion Mass Spectrometry SIMS technique to obtain micrographs of the elemental (or isotopic) distribution at the surface of a sample with a spatial resolution of 2 mm or better.

### Ion

An atom or a molecule that has lost or gained an electron. An ion is also known as a charged particle. Ions are usually more reactive than neutral, uncharged particles. They cause all sorts of things from static electricity and chemical reactions like smog, to body processes like cellular respiration and digestion.

### Ionophore

A macro-organic molecule capable of specifically solubilizing an inorganic ion of suitable size in organic mediums.

### Ionosonde

An instrument used to measure the *ionization* of gases in the atmosphere of a planet or a moon.

### Ionosphere

A region of gases, very high up in the *atmosphere,* that contains *ionized* gases.

### ISE (Ion Selective Electrode)

Ions in solution are quantified by measuring the change in voltage (i.e. potentiometric) resulting from the distribution of ions (by ion exchange controlled by the ion exchange current io) between a sensing membrane (the ion selective membrane) and the solution. This potential is measured at zero current with respect to a reference electrode which is also in contact with the solution. The potential measured is proportional to the logarithm of the analyte concentration. The oldest and best known ISE is the pH sensor based on a glass membrane. More recently, polymeric membranes have been formed incorporating ionophores rendering the membrane specific to certain ions only.

## Isoelectronic

Two molecules are described as isoelectronic if they have the same number of valence electrons in similar orbitals, although they may differ in their distribution of nuclear charges (e.g., H-CN and $H\text{-}N^{+}C^{-}$. KineticPertaining to the rates of chemical reactions. A fast reaction is said to have fast kinetics; if the balance of products in a reaction is controlled by reaction rates rather than by thermodynamic equilibria, the reaction is said to be kinetically controlled.

## Isothermal

An *isothermal process* is a thermodynamic process in which the temperature of the system stays constant: "$T = 0$. This typically occurs when a system is in contact with an outside thermal reservoir (heat bath), and processes occur slowly enough to allow the system to continually adjust to the temperature of the reservoir through heat exchange. An alternative special case in which a system exchanges no heat with its surroundings ($Q = 0$) is called an *adiabatic process*.

# J

### Junna

Minor Negentropy Alliance world. During the period from 3267 to 3421 was center for the Institute of Baseline Psychotypology. Suffered during the Version War and became an MPA garrison. Since the age of separative empires the locals have drifted increasingly towards Zoeticism

### Jupiter-brain

An AI, posthuman or Local Archailect Node of extremely high computational power and size.

This is the typical megascale concentrated intelligence.

The term originated due to an idea by early Information Age transhumanist Keith Henson that nanomachines could be used to turn the mass of Jupiter into computers running an upgraded version of himself.

# K

## Keratin

Keratins are a family of fibrous structural proteins; tough and insoluble, they form the hard but nonmineralized structures found in reptiles, birds, amphibians and mammals. The baleen plates of filter-feeding whales are made of them, as are human fingernails. Keratins are also found in the gastrointestinal tracts of many animals, including roundworms. They are rivaled in biological toughness only by chitin, a cellulose-like polymer of glucosamine and the main constituent of the exoskeletons of arthropods. There are various types of keratins, even within a single animal. Some infectious fungi, such as those which cause athlete's foot and ringworm, feed on keratin. The silk fibroins produced by insects and spiders are often classified as keratins, though it is unclear whether they are phylogenetically related to vertebrate keratins.

Cells in the epidermis contain a structural matrix of keratin which makes this outermost layer of the skin almost waterproof, and along with collagen and elastin, gives skin its strength. Rubbing and pressure cause keratin to proliferate with the formation of protective calluses — useful for athletes and on the fingertips of musicians who play stringed instruments. Keratinized epidermal cells are constantly shed and replaced.

In mammals there are soft epithelial keratins, the cytokeratins, and harder hair keratins. As certain skin cells differentiate and become cornified, pre-keratin polypeptides are incorporated into intermediate filaments. Eventually the nucleus and cytoplasmic organelles disappear, metabolism ceases and cells undergo a programmed death as they become fully keratinized.

Keratins are the main constituent of structures that grow from the skin: the ±-keratins in the hair (including wool), horns, nails, claws and hooves of mammals and the harder $^{2}$-keratins in the scales and claws of reptiles, their shells (chelonians, such as tortoise, turtle, terrapin), and in the feathers, beaks, and claws of birds. Although it is now difficult to be certain, the scales, claws, some protective armour and the beaks of dinosaurs would, almost certainly, have been composed of a type of keratin. In Crossopterygian fish, the outer layer of cosmoid scales was keratin.

These hard, integumentary structures are formed by intercellular cementing of fibers formed from the dead, cornified cells generated by specialized beds deep within the skin. Hair grows continuously and feathers moult and regenerate. The constituent proteins may be phylogenetically homologous but differ somewhat in chemical structure and supermolecular organization. The evolutionary relationships are complex and only partially known. Multiple genes have been identified for the $^{2}$-keratins in feathers, and this is probably characteristic of all keratins.

## Ketone

A ketone is either the functional group characterized by a carbonyl group linked to two other carbon atoms or a chemical compound that contains this functional group. A ketone can be generally represented by the formula:

$$R_1(CO)R_2.$$

A carbonyl carbon bonded to two carbon atoms distinguishes ketones from carboxylic acids, aldehydes, esters, amides, and other oxygen-containing compounds. The double-bond of the carbonyl group distinguishes ketones from alcohols and ethers. The simplest ketone is acetone (also called propanone).

The carbon atom adjacent to a carbonyl group is called the ±-carbon. Hydrogens attached to this carbon are called ± hydrogens. In the presence of an acid catalyst the ketone is subjected to so-called keto-enol tautomerism. The reaction with a strong base gives the corresponding enolate. A diketone is a compound

*Nomenclature*

In general, ketones are named using IUPAC nomenclature by changing the suffix *-e* of the parent alkane to *-one*. For common ketones, some traditional names such as acetone and benzophenone predominate, and these are considered retained IUPAC names, although some introductory chemistry texts use names such as 2-propanone or propanone.

Oxo is the formal IUPAC nomenclature for a ketone functional group. However, other prefixes are also used by various books and journals. For some common chemicals (mainly in biochemistry), keto or oxy is the term used to describe the ketone (also known as alkanone) functional group. Oxo also refers to a single oxygen atom coordinated to a transition metal (a metal oxo).

*Physical properties*

A carbonyl group is polar. This makes ketones polar compounds. The carbonyl groups interact with water by hydrogen bonding. It is a hydrogen-bond acceptor, but not a hydrogen-bond donator, and cannot hydrogen-bond to itself. This makes ketones more volatile than alcohols and carboxylic acids of similar molecular weight.

*Acidity*

The ±-hydrogen of a ketone is far more acidic (pKa H″ 20) than the hydrogen of a regular alkane (pKa H″ 50). This is due to resonance stabilization of the enolate ion that is formed through dissociation. The relative acidity of the ±-hydrogen is important in the enolization reactions of ketones and other carbonyl compounds.

*Spectroscopic properties*

Spectroscopy is an important means for identifying ketones. Ketones and aldehydes will display a significant peak in infrared spectroscopy, at around 1700 cm″1 (slightly higher or lower, depending on the chemical environment)

## Kevlar

A synthetic fiber made by E. I. du Pont de Nemours & Co., Inc. Stronger than most steels, Kevlar is among the strongest commercially available materials and is used in aerospace construction, bulletproof vests, and other applications requiring a high strength-to-weight ratio.

## Kilocalorie (kcal)

Unit of heat energy equal to 1000 calories. Often used to express the energy content of food or molecules: bond strengths, for example, are measured in kcal/mole. An alternative unit in wide use is the kilojoule, equal to 0.24 kcal.

## Kinesin

Kinesin is a class of motor protein dimer found in biological cells. A kinesin attaches to microtubules, and moves along the tubule in order to transport cellular cargo, such as vesicles. Kinesins typically consist of two large globular heads that allow attachment to microtubules, a central coiled region, and a region termed light-chain, which connects the kinesin to the intracellular component to be moved. Most kinesin-related proteins move toward the plus end of the microtubule, as does kinesin itself, but some move toward the minus end.

Kinesin and Dynein lie at the heart of Microtubule-Based Movement. These are termed Microtubule Associated Proteins (MAPs). These motor MAPs attach both to intracellular components, and to microtubles (MTs), and by moving along the MT they are able to transport the intracellular components, which could be organelles, vesicles, or other components of the cytoskeleton, to where they are required.

Kinesin accomplishes transport by essentially "walking" along a microtubule. Two mechanisms were proposed to explain how this movement occurs. In the "hand-over-hand" mechanism, the kinesin heads step over one another, in effect alternating in the role of leading forward. In the "inch-worm" mechanism, one kinesin dimer moves forward, with the other one then being dragged forward. Despite some remaining controversy, the mounting evidence points towards the hand-over-hand mechanism as being more likely.

In recent years, it has been found that microtubule-based molecular motors including a number of kinesin motors, are capable of organizing two separate microtubule asters into a metastable structures independent of any external positional cues. This self-organization is in turn dependent on the directionalities of these molecular motors as well as its processivity. It is also dependent on the dynamic instability and stochastic transition of microtubule aster itself. This motor-dependent self-organization ultimately contributed to a centrsome-independent mitotic spindle assembly mechanism. While many kinesins walk along microtubules toward either the plus or minus ends, the Kinesin 13 family act as regulators of microtubule dynamics. The protypical member of this family is MCAK (formerly $Kif_2C$, XKCM1) which acts at the ends of microtubule polymers to depolymerize them. The function of MCAK in cells and its mechanism in vitro is currently being investigated by numerous labs.

## Kinetic Energy

Kinetic energy (SI unit: the joule) is energy that a body possesses as a result of its motion. It is formally defined as *the work needed to accelerate a body from rest to its current velocity*. Having gained this energy during its acceleration, the body maintains this kinetic energy unless its speed changes. Negative work of the same magnitude would be required to return the body to a state of rest from that velocity.

## Kinetic Theory

Kinetic theory attempts to explain macroscopic properties of gases, such as pressure, temperature, or volume, by considering their molecular composition and motion. Essentially, the theory posits that pressure is due not to static repulsion between molecule, as was Isaac Newton's conjecture, but due to collisions between molecules moving about with a certain velocity. Kinetic theory is also known as *kinetic-molecular theory* or *collision theory*. In 1738, Swiss physician and mathematician Daniel Bernoulli published *Hydrodynamica* which laid the basis for the kinetic theory of gases. In this work, Bernoulli positioned the argument, still used to this day, that gases consist of great numbers of molecules moving in all directions, that their impact on a surface causes the gas pressure that we feel, and that what we experience as heat is simply the kinetic energy of their motion. In 1859, after reading a paper on the diffusion of molecules by Rudolf Clausius, Scottish physicist James Clerk Maxwell formulated the Maxwell distribution of molecular velocities, which gave the proportion of molecules having a certain velocity in a specific range. This was the first-ever statistical law in physics

## Knowbots

A *knowbot* is a kind of bot that collects information by automatically gathering certain specified information from web sites.

# L

## Lab-on-a-chip

This term for devices that integrate (multiple) laboratory functions on a single chip of only millimeters to a few square centimeters in size and that are capable of handling extremely small fluid volumes down to less than pico liters. Lab-on-a-chip devices are a subset of MEMS devices and often indicated by "Micro Total Analysis Systems" (µTAS) as well. Microfluidics is a broader term that describes also mechanical flow control devices like pumps and valves or sensors like flowmeters and viscometers. However, strictly regarded "Lab-on-a-Chip" indicates generally the scaling of single or multiple lab processes down to chip-format, whereas "µTAS" is dedicated to the integration of the total sequence of lab processes to perform chemical analysis. The term "Lab-on-a-Chip" was introduced later on when it turned out that µTAS technologies were more widely applicable than only for analysis purposes. After the discovery of microtechnology (~1958) for realizing integrated semiconductor structures for microelectronic chips, these lithography-based technologies were soon applied in pressure sensor manufacturing (1966) as well.

Due to further development of these usually CMOS-compatibility limited processes, a tool box became available to create micrometre or sub-micrometre sized mechanical structures in silicon wafers as well: the Micro Electro Mechanical Systems (MEMS) era (also indicated with Micro System Technology - MST) had started. Next to pressure sensors, airbag sensors and other mechanically movable structures, fluid handling devices were developed. Examples are: channels (capillary connections), mixers, valves, pumps and dosing devices. The first LOC analysis system was a gas chromatograph, developed in 1975 by S.C. Terry - Stanford University. However, only at the end of the 1980's, and beginning of the 1990's, the LOC research started to seriously grow as a few research groups in Europe developed micropumps, flowsensors and the concepts for integrated fluid treatments for analysis systems. These µTAS concepts demonstrated that integration of pretreatment steps, usually done at lab-scale, could extend the simple sensor functionality towards a complete laboratory analysis, including e.g. additional cleaning and separation steps. A big boost in research and commercial interest came in the mid 1990's, when µTAS technologies turned out to provide interesting tooling for genomics applications, like capillary electrophoresis and DNA microarrays.

A big boost in research support also came from the military, especially from DARPA (Defense Advanced Research Projects Agency), for their interest in portable bio/chemical warfare agent detection systems. The added value was not only limited to integration of lab processes for analysis but also the characteristic possibilities of individual components and the application to other, non-analysis, lab processes. Hence the term "Lab-on-a-Chip" was introduced. Although the application of LOCs is still novel and modest, a growing interest of companies and applied research groups is observed in different fields such as analysis (e.g. chemical analysis, environmental monitoring, medical diagnostics and cellomics) but also in synthetic chemistry (e.g. rapid screening and microreactors for pharmaceutics). Besides further application developments, research in LOC systems is expected to extend towards downscaling of fluid handling structures as well, by using nanotechnology.

Sub-micrometre and nano-sized channels, DNA labyrinths, single cell detection an analysis and nano-sensors might become feasible that allow new ways of interaction with biological species and large molecules.

The basis for most LOC fabrication processes is lithography. Initially most processes were in silicon, as these well-developed technologies were directly derived from semiconductor fabrication. Because of demands for e.g. specific optical characteristics, bio- or chemical compatibility, lower production costs and faster prototyping, new processes have been developed such as glass, ceramics and metal etching, deposition and bonding, PDMS processing (e.g., soft lithography), thick-film- and stereolithography as well as fast replication methods via electroplating, injection molding and embossing. Furthermore the LOC field more and more exceeds the borders between lithography-based microsystem technology, nano technology and precision engineering.

## Lacrosse

A spy satellite launched by NASA in December of 1988. Lacrosse's main instrument, like most spy satellites, was its image sensor. It also carried *SAR*.

## Lagging Strand

In DNA replication, the lagging strand is the DNA strand at the opposite side of the replication fork from the leading strand. It goes from 5' to 3' (these numbers indicate the position of the molecule in respect to the carbon atoms it contains).

When replicating, the original DNA splits in two, forming two "prongs" which resemble a fork (i.e. the "replication fork"). DNA has a ladder-like structure; imagine a ladder broken in half vertically, along the steps. Each half of the ladder now requires a new half to match it.

Pol III, the main DNA replication enzyme, can not work in the 5' - 3' direction, and so replication of the lagging strand is more complicated than of the leading strand. On the leading strand, Pol III "reads" the DNA and adds nucleotides to it continuously.

On the lagging strand, primase "reads" the DNA and adds RNA to it in short bursts. Pol III lengthens the bursts, forming Okazaki fragments. Pol I then "reads" the fragments, removes the RNA, and adds its own nucleotides (this is necessary because RNA and DNA use slightly different kinds of nucleotides). DNA ligase joins the fragments together.

### Lamins

Intermediate filament proteins that form the nuclear lamina on the inner surface of the nuclear envelope.

### Langmuir-Blodgett

The name of a nanofabrication technique used to create ultrathin films (monolayers and isolated molecular layers), the end result of which is called a "Langmuir-Blodgett film".

### Laser Trimming

Laser trimming is a term that describes the manufacturing process of using a LASER to adjust the operating parameters of an electronic circuit.

The usual approach is to use a laser to burn away small portions of resistors, raising their value (resistance). The burning operation can be conducted while the circuit is being tested by automatic test equipment, leading to extremely accurate (appropriate) final values for the trimming resistor(s) (also known as AOT - Adjust on test).

Laser trimming is the controlled alteration of the attributes of a capacitor or a resistor by a laser beam. Selecting one or more components on the circuit and adjusting them with the laser achieves this. The trim changes the resistor or capacitor value until the nominal value has been reached.

The resistance value of a film resistor is defined by his geometric dimensions (length, width, height) and the resistor material. A lateral cut in the resistor material by the laser narrows the current flow path and increases the resistance value. The same effect is obtained whether the laser changes a thick-film or a thin-film resistor on a ceramic substrate or an SMD-resistor on a SMD circuit. The SMD-resistor is produced with the same technology and normally is laser trimmed as well.

Trimmable chip capacitors are build up as multilayer plate capacitors. Vaporizing the top layer with a laser decreases the capacitance by reducing the area of the top electrode.

Passive trim is the adjustment of a resistor to a given value. If the trimming adjusts the whole circuit output (e.g. output voltage, frequency, switching threshold ...), this is called active trim. During the trim process, the corresponding parameter is measured continuously and compared to the programmed nominal value. The laser stops automatically when the value reaches the nominal value.

Often designers use potentiometers, which would have to be adjusted during end testing until the desired function of the circuit has been reached. In many applications, the end user of the products does not allow potentiometers. Therefore manufacturers determine the needed resistance or capacitance values by measurement and calculation methods and afterwards solder the suitable component into the final PCB.

It is simpler to substitute the potentiometer or the adjust element with a trimable chip resistor or chip capacitor and the potentiometer adjusting screwdriver with the laser and active trimming. The achieved accuracy is higher, the procedure can be automated and the long term stability is better than at potentiometers or at least in the same region as replace-and-resolder chip components. Often the laser for the active trim could can integrated in existing measurement places at the customer factories.

A similar approach can be used to program digital logic circuits. In this case, fuses are blown by the laser, enabling or disabling various logic circuits. An example of this is the IBM Power-4 processor chip where the chip contains five banks of cache memory but only requires four banks for full operation. During testing, each cache bank is exercised. If a defect is found in one bank, that bank can be disabled by blowing its programming fuse. This built-in redundancy allows higher chip yields than would be possible if all cache banks had to be perfect in every chip. (If no bank is defective, a fuse can be blown arbitrarily, leaving just four banks.)

## LDEF

NASA's Long Duration Exposure Facility (LDEF) was launched in 1984 to study the effect of spending a long time exposed to the harsh conditions of space on different materials. In orbit for over five years, LDEF carried 57 experiments belonging to more than 200 investigators from universities, private companies, and NASA centres.

### Leakage

The loss of all or parts of a useful agent, as of the electric current that flows through an insulator or the magnetic flux that passes outside useful flux circuits.

### Lectin

Protein that binds tightly to a specific sugar. Abundant lectins derived from plant seeds are often used as affinity reagents to purify glycoproteins or to detect them on the surface of cells.

### Leucine Zipper

Structural motif seen in many DNA-binding proteins in which two a helices from separate proteins are joined together in a coiled-coil, forming a protein dimer.

### LIF Laser Induced Fluorescence

The optical emission from molecules that have been excited to higher energy levels by absorption of electromagnetic radiation. The main advantage of fluorescence detection compared to absorption measurements is the greater sensitivity achievable because the fluorescence signal has a very low background. For molecules that can be resonant excitated, LIF provides selective excitation of the analyte to avoid interferences. LIF is useful to study the electronic structure of molecules and to make quantitative measurements of analyte concentrations. Analytical applications include monitoring gas-phase concentrations in the atmosphere, flames, and plasmas; and remote sensing using light detection and ranging (LIDAR).

### Ligand

In protein chemistry, a small molecule that is (or can be) bound by a larger molecule is termed a ligand. In organometallic chemistry, a moiety bonded to a central metal atom is also termed a ligand; the latter definition is more common in general chemistry.

## Ligand-directed Drug

The first goal is to select a small number of likely targets (which at least in some cases, such as targets in pathogens, are validated targets) rapidly, without the need for costly and time- consuming biologically based target validation and functional determination. The second goal is to obtain lead compounds for these targets, instead of starting with a biologically validated target and then attempting to screen libraries for lead compounds that modulate it.

## Light Emitting Diodes

Light Emitting Diodes (LEDs) work on a completely different concept. Traditionally LEDs are created from two semiconductors. By running current in one direction across the semiconductor the LED emits light of a particular frequency (hence a particular color) depending on the physical characteristics of the semiconductor used. The semiconductor is covered with a piece of plastic that focuses the light and increases the brightness. These semiconductors are very durable, there is no filament, they donít require much power, theyíre brighter and they last a long time. By densely packing red, blue and green LEDs next to each other on a substrate one can create a display.

## Lightsail

A spacecraft propulsion system that gains thrust from the pressure of light striking a thin metal film.

## Limit of Detection

The smallest measurable input. This differs from resolution, which defines the smallest measurable change in input. For a temperature measurement, this would provide an indication of the lowest temperature a sensor could generate an output in response to.

## Limited Assembler

An assembler with built-in limits that constrain its use (for example, to make hazardous uses difficult or impossible, or to build just one thing). Assembler capable of making only certain products; faster, more efficient, and less liable to abuse than a general-purpose assembler.

## Linde Scenario

A scenario for indefinite survival of intelligent life. It assumes it is possible to either create basement universes connected to the original universe with a wormhole *or* the existence of other cosmological domains. Intelligent life continually migrates to the new domains as the old grow too entropic to sustain life.

## Linear

Aside from its geometric meaning, linear describes systems in which an output is directly proportional to an input. In particular, a linear elastic system is one in which the internal displacements are (at equilibrium) directly proportional to applied forces.

## Liposome

A type of nanoparticle made of lipids, or fat molecules, surrounding a water core. Liposomes, several of which are widely used to treat infectious diseases and cancer, were the first type of nanoparticle to be used to create therapeutic agents with novel characteristics.

## Liquid Crystal Display

Liquid Crystal Display — display device employing light source and electrically alterable optically active thin film.

## Lofstrom Loop

An beanstalk-like megaconstruction based on a stream of magnetically accelerated bars linked together. The stream is sent into space, where a station rides it using magnetic hooks, redirects it horizontally to another station, which sends it downwards to a receiving station on the ground. From this station the stream is then sent back to the launch station (a purely vertical version is called a space fountain). This structure would contain a large amount of kinetic energy but could be built gradually and would only require enough energy to compensate for losses when finished. Elevators could be run along the streams, and geostationary installations could be placed along the horizontal top.

### London Dispersion Force

An attractive force between atoms or molecules caused by the numerous transient dipoles resulting from electronic superposition.

### Lone Pair

A lone pair is an electron pair without bonding or sharing with other atoms. It is found in the outermo$_{s}$t electron shell of an atom. It often exhibits a negative polar character with its high charge density. It is used in the formation of a dative bond, for example, the creation of the hydroxonium, H3O+, ion occurs when acids are dissolved in water and it is due to the oxygen atom donating a lone pair to the hydrogen ion.

### Low Earth Orbit

An orbit within the Earth's atmosphere, but at its highest layer. Any satellite in a low Earth orbit can make observations of the Earth from fairly low down.

# M

## Macromolecule

A macromolecule is a molecule with a large molecular mass, but generally the use of the term is restricted to polymers and molecules which structurally include polymers. Many examples come from biology and in particular biochemistry. In case of "biomacromolecules" or biopolymers, there are proteins, carbohydrates, nucleic acids (such as DNA), and lipids (fat). Synthetic examples include plastics. The integral domains of crystals and metals, while composed of very large numbers of atoms joined by molecule-like bonds, are rarely referred to as "macromolecules."

The term macromolecule is also sometimes used to refer to aggregates of two or more macromolecules held together by intermolecular forces rather than by chemical "bonds". This usage is common in particular when the individual macromolecules involved aggregate or "assemble" spontaneously and rarely exist in isolation. Such an aggregate is more properly called a macromolecular complex. In such a context, individual macromolecules are often referred to as subunits ).

Substances that are composed of macromolecules often have unusual physical properties. The properties of liquid crystals and such elastomers as rubber are examples. Although too small to *see, individual pie*ces of DNA in solution can be broken in two simply by suctioning the solution through an ordinary straw. This is not true of smaller molecules. The 1964 edition of Linus Pauling's College Chemistry asserted that DNA in nature is never longer than about

5000 base pairs. This is because biochemists were inadvertently and with perfect consistency breaking their samples into pieces. In fact, the DNA of chromosomes can be tens of millions of base pairs long.

Another common macromolecular property that does not characterize smaller molecules is the need for assistance in dissolving into solution. Many require salts or particular ions to dissolve in water. Proteins will denature if the solute concentration of their solution is too high or too low. According to IUPAC recommendations the term macromolecule is reserved for an individual molecule, and the term polymer is used as to denote a substance composed of macromolecules. Polymer may also be employed unambiguously as an adjective, according to accepted usage, e.g. polymer blend, polymer molecule.

## Macroscale

Larger than nanoscale; often implies a design that humans can directly interact with; too large to be built by a single assembler (one cubic micron of diamond contains 176 billion atoms).

## Macrosensing

In me*dical nanorobotics, the detecti*on of global somatic states (inside the human body) and extrasomatic states (sensory data originating outside of the human body) by in vivo nanorobots.

## Magnetic Force Microscopy

A method for observing local magnetic fields near a surface by scanning the surface with a magnetic probe.

## Magnetic Moment

Elementary particles possess an intrinsic quantum mechanical property known as spin. This is analogous to the angular momentum of an object that is spinning around its center of mass, although strictly speaking these particles are believed to be point-like and cannot be said to be rotating. Spin is measured in units of the reduced Planck constant (), with electrons, protons and neutrons all having spin ½ , or "spin-½". In an atom, electrons in motion around the nucleus possess orbital angular momentum in addition to their spin, while the nucleus itself possesses angular momentum due to its nuclear spin.

The magnetic field produced by an atom—its magnetic moment—is determined by these various forms of angular momentum, just as a rotating charged object classically produces a magnetic field. However, the most dominant contribution comes from spin. Due to the nature of electrons to obey the Pauli exclusion principle, in which no two electrons may be found in the same quantum state, bound electrons pair up with each other, with one member of each pair in a spin up state and the other in the opposite, spin down state. Thus these spins cancel each other out, reducing the total magnetic dipole moment to zero in some atoms with even number of electrons.

In ferromagnetic elements such as iron, an odd number of electrons leads to an unpaired electron and a net overall magnetic moment. The orbitals of neighboring atoms overlap and a lower energy state is achieved when the spins of unpaired electrons are aligned with each other, a process known as an exchange interaction. When the magnetic moments of ferromagnetic atoms are lined up, the material can produce a measurable macroscopic field. Paramagnetic materials have atoms with magnetic moments that line up in random directions when no magnetic field is present, but the magnetic moments of the individual atoms line up in the presence of a field. The nucleus of an atom can also have a net spin. Normally these nuclei are aligned in random directions because of thermal equilibrium. However, for certain elements (such as xenon-129) it is possible to polarize a significant proportion of the nuclear spin states so that they are aligned in the same direction—a condition called hyperpolarization. This has important applications in magnetic resonance imaging.

## Magnetic Resonance Force Microscopy MRFM

A new microscopic imaging technique that combines aspects of atomic force microscopy AFM with magnetic resonance imaging MRI. Interest in this technique is driven by the possibility of reaching the "Holy Grail" of microscopy: true three-dimensional, sub- surface imaging with atomic resolution and chemical specificity.

## Magnetic Resonance Imaging

MRI or Magnetic Resonance Imaging is a relatively new technology that is revolutionizing several fields. This method of scanning was developed primarily for use in medicine but it has also been used to study fossils and historical artefacts. Early doctors were only able to gather data about a patient through observation and rudimentary tests. X-Rays provided doctors with one of the first ways of peering withi*n a li*ving person. The MRI is one of the exciting successors to the X-Ray.

To perform a MRI scan, the patient is securely placed on an imaging table within a large MRI scanner. Powerful magnetic fields are administered to align the nuclei within the atoms of the patient's body. Next, radio frequency pulses are applied; finally, the nuclei release some of the radio frequency energy and these emissions are detected by the MRI equipment. With this data, a computer generates a surprisingly detailed view of tissues within the body.

Earlier imaging technologies, such as X-rays, were able to detect dense tissues, particularly bones. MRIs give doctors the ability to view all sorts of body structures including soft tissues.

MRIs are frequently used to detect cancers that would otherwise be difficult to diagnose, such as *mesothelioma*. The ability to detect cancers at their early stages has brought these scanners to the forefront of the battle against many diseases. It is generally believed that patients are not harmed by undergoing the procedure since MRIs do not use radiation.

There are not any side-effects, but patients with pacemakers or other metallic implants are not eligible for these scans. Exams typically take between 30 minutes and one hour. Early models of MRI scanners required patients to be placed in confined positions; newer versions of these expensive machines, however, are based on an open design that is much more spacious and comfortable.

## Magnetic Resonance Microscopy

Confocal or optical microscopy (OM) and magnetic resonance microscopy (MRM) have developed as important tools for cellular research. MRM is noninvasive and nondestructive, and OM requires only the expression or uptake of fluorescently labeled molecules for detection. Both methods have their advantages and disadvantages. MRM provides access to several observable quantities that cannot be determined with OM alone (e.g., metabolite concentrations, chemical shifts, spin couplings, T1 and T2 relaxation times, and diffusion constants). These quantities have been related to a variety of such cellular events as tumor formation, programmed death (apoptosis), necrosis, and increased proliferation. Instruments combining OM and MRM allow live cells to be studied simultaneously using both techniques, providing a necessary link between cellular response and molecular information on proteins and other biochemicals involved in a certain cellular event.

## Magnetization (M)

The total magnetic moment per unit volume of material. Also, a measure of the contribution to the magnetic flux by some material within an H field. The magnitude of M is proportional to the applied field as: $M = \div_m \times H$, with $\div_m$ the magnetic susceptibility.

## Magnetosphere

The space around the Earth in which *ions* (charged particles) are controlled by the Earth's own magnetic field.

## Magnetostrictive Material

A material that changes dimension in the presence of a magnetic field or generates a magnetic field when mechanically deformed.

## Manobarcodes Particles

Nanobarcodes™ particles, comprise freestanding, cylindrically- shaped metal nanoparticles that are self- encoded with sub- micron stripes. Nanobarcodes particles are prepared by electrochemical reduction of metal ions into the cylindrical pores of metallized templates, followed by release from the template. Intrinsic differences in reflectivity between adjacent metal stripes allow individual particles to be identified by conventional optical microscopy.

By varying the physical dimensions (length and width), the composition (number/type of metals) and striping pattern (number/ width/ order of stripes), libraries of thousands of uniquely identifiable particle types ("flavors") can be prepared. Nanobarcodes particles are thus the nanoscale equivalent of conventional bar codes.

### Mars Observer

A NASA *space probe* launched September 1992 to study Mars.

### Martensite

A metastable iron phase supersaturated in carbon that is the product of a diffusionless (athermal) transformation from austenite.

### Mask

Pattern on glass, like a photographic negative, for producing integrated-circuit elements on semiconductor wafer.

### Mass Spectrometry

This technique can be used to both measure and analyze molecules under study. It involves introducing enough energy into a target molecule to cause its ionization and disintegration. The resulting fragments are then analyzed, based on the mass/ charge ratio to produce a "molecular fingerprint."

### Mass to Charge Ratio

A number defining how a particle will respond to an electric or magnetic field that can be calculated by dividing the mass of a particle by its charge. For instance, mass spectrometry provides information on the mass to charge ratio of a molecular species. In this case, the ratio is expressed by m/z where m is the molecular weight and z is the elementary charge.

### Massometer

In medical nanorobotics, a nanosensor device for measuring the mass of individual molecules or small physical objects to single-proton resolution.

### Materials Science

Science of ceramics, glass,

metals, plastics, semiconductors.

## MatML Materials Markup Language

Materials property data distributed on the World Wide Web in documents using hypertext markup language.

## Mechanochemistry

. Mechanochemistry, sometimes also called "positional synthesis" or "positional assembly" is a technique for forming chemical bonds by direct computer control of the position of molecules. Mechanochemistry is a specific type of Mechanosynthesis.

As of 2004, the typical experimental arrangement is to attach a molecule to the tip of an atomic force microscope, and then use the microscope's precise positioning abilities to push the molecule on the tip into another on a substrate. Since the angles and distances can be precisely controlled, and the reaction occurs in a vacuum, novel chemical compounds and arrangements are possible. Much of the excitement regarding mechanochemistry regards its potential use in automated assembly of molecular-scale devices. Such techniques appear to have many applications in medicine, aviation, resource extraction, manufacturing and warfare.

Most theoretical explorations of such machines have focused on using Carbon, because of the many strong bonds it can form, the many types of chemistry these bonds permit, and utility of these bonds in medical and mechanical applications. Carbon forms diamond, for example, which if cheaply available, would be an excellent material for many machines.

In practice, getting exactly one molecule to a known place on the microscope's tip is possible, but has proven difficult to automate. Since practical products require at least several hundred million atoms, this technique has not yet proven practical in forming a real product.

The goal of mechano-assembly research at this point focuses on overcoming these problems by calibration, and selection of appropriate synthesis reactions. The first product to be built by these means will probably be a specialized, very small (roughly 1,000 nanometers on a side) machine tool that can build copies of itself using mechanochemical means, under the control of an external computer. In the literature, such a tool is called an assembler. Once assemblers exist, geometric growth (copies making copies) could reduce the cost of assemblers rapidly. Control by an external computer should then permit large groups of assemblers to construct large, useful projects to atomic precisions. One such project would combine molecular-level conveyor belts with permanently-mounted assemblers to produce a factory.

## Mechanosynthesis

In conventional chemical synthesis or chemosynthesis, reactive molecules encounter one another through random thermal motion in a liquid or vapor. In a hypothesized process of mechanosynthesis, reactive molecules would be attached to molecular mechanical systems, and their encounters would result from mechanical motions bringing them together in planned sequences, positions, and orientations. It is envisioned that mechanosynthesis would avoid unwanted reactions by keeping potential reactants apart, and would strongly favor desired reactions by holding reactants together in optimal orientations for many molecular vibration cycles. Mechanosynthetic systems would be designed to resemble some biological mechanisms.

While the description of mechanosynthesis given above has not yet been achieved, primitve mechanochemistry has been performed at cryogenic temperatures using scanning tunneling scraping electron microscopes). So far, such devices provide the closest approach to fabrication tools for molecular engineering.

Broader exploitation of mechanosynthesis awaits more advanced technology for constructing molecular machine systems - including a molecular assembler or precursors thereof.

It has been suggested, notably by K. Eric Drexler, that mechanosynthesis will be fundamental to molecular manufacturing based on nanofactories capable of building macroscopic objects with atomic precision. The potential for these has been disputed, notably by Nobel Laurate Richard Smalley, leading to a famous dispute between the two of them .

In part to resolve this and related questions about the dangers of industrial accidents and runaway events equivalent to Chernobyl and Bhopal, and the more remote issue of ecophagy, grey goo and green goo (various potential disasters arising from runaway replicators, which could be built using mechanosynthesis) the UK Royal Society and UK Royal Academy of Engineering in 2003 commissioned a study to deal with these issues and larger social and ecological implications, led by mechanical engineering professor Ann Dowling. This was anticipated by some to take a strong position on these problems and potentials - and suggest any development path to a general theory of so-called mechanosynthesis.

However, the Royal Society's nanotech report did not address molecular manufacturing at all, except to dismiss it along with gray goo.

### Mechatronics

The synergistic combination of precision mechanical engineering with electronic control.

### Megabyte (MB)

$2^{20}$ (= 1,048,576, or about one million) bytes of information.

### Meme

An idea that replicates through a society as it is propagated through person-to-person interaction, both direct and indirect. Memetics is a field of study that focuses on memes' role in the evolution of a culture. Self-reproducing idea or other information pattern which is propagated in ways similar to that of a gene.

### MEMS MicroElectro Mechanical Systems

MEMS is less an application area in itself than a manufacturing or fabrication technique that enables other application areas. Many authors use MEMS as shorthand to imply a number of particular application areas. As it is used here, MEMS is a "top- down" fabrication technology that is especially useful for integrating mechanical and electrical systems together on the same chip. It is grouped in the category of integrated microsystems because these same MEMS techniques can be extended in the future to also help integrate biological and chemical components on the same chip, as discussed below. Thus far, MEMS techniques have been used to make some functional commercial devices such as sensors and single-chip measurement devices. Many researchers have used MEMS technologies as analytical tools in other areas of nanotechnology.

Stood originally for Micro-ElectroMechanical System — microscopic mechanical elements, fabricated on silicon chips by techniques similar to those used in integrated circuit manufacture, for use as sensors, actuators, and other devices. Today almost any miniaturized device (based on Si technology or traditional precision engineering, chemical or mechanical) is referred to as a MEMS device generic term to describe micron scale electrical/mechanical devices.

## Meridian Lines

The imaginary lines that were drawn on the globe for navigational purposes. Meridian lines mark longitude and run through both the north and the south poles. They separate the Earth into time zones and they are used to indicate a coordinate on the Earth. The Greenwich meridian is the central meridian line from which all time zones are set.

## Mesh Networking

Mesh networking is a way to route data, voice and instructions between nodes. It allows for continuous connections and reconfiguration around blocked paths by "hopping" from node to node until a connection can be established.

Mesh networks are self-healing: the network can still operate even when a node breaks down or a connection goes bad. As a result, a very reliable network is formed. This concept is applicable to wireless networks, wired networks, and software interaction.

A mesh network is a networking technique which allows inexpensive peer network nodes to supply back haul services to other nodes in the same network. It effectively extends a network by sharing access to higher cost network infrastructure.

Mesh networks differ from other networks in that the component parts can all connect to each other.

An MIT project developing "hundred dollar laptops" for under-privileged schools in developing nations plans to use mesh networking to create a robust and inexpensive infrastructure for the students who will receive the laptops. The instantaneous connections made by the laptops is claimed by the project to reduce the need for an external infrastructure such as the internet to reach all areas, because a connected node could share the connection with nodes nearby.

In Cambridge, UK, on the 3rd June 2006, mesh networking was used at the "Strawberry Fair" to run mobile live television, radio and internet services to an estimated 80,000 people.

## Mesoscale

A device or structure larger than the nanoscale (10^-9 m) and smaller than the megascale; the exact size depends heavily on the context and usually ranges between very large nanodevices (10^-7 m) and the human scale (1 m).

## Messenger Molecule

A chemically recognizable molecule which can convey information after it is received and decoded by an appropriate chemical sensor.

## Metal Nanoshells

A new type of nanoparticle composed of a semiconductor or dielectric core coated with an ultrathin conductive layer.. By adjusting the relative core and shell thicknesses, metal nanoshells can be fabricated that will absorb or scatter light at any wavelength across the entire visible and infrared range of the electromagnetic spectrum.

## Metabolism

In contrast to higher organisms, bacteria exhibit an extremely wide variety of metabolic types. In fact, it is widely accepted that eukaryotic metabolism is largely a derivative of bacterial metabolism with mitochondria having descended from a lineage within the ±-Proteobacteria and chloroplasts from the Cyanobacteria by ancient endosymbiotic events. Bacterial metabolism can be divided broadly on the basis of the kind of energy used for growth, electron donors and electron acceptors and by the source of carbon used. Most bacteria are heterotrophic; using organic carbon compounds as both carbon and energy sources. In aerobic organisms, oxygen is used as the terminal electron acceptor.

In anaerobic organisms other inorganic compounds, such as nitrate, sulfate or carbon dioxide as terminal electron acceptors leading to the environmentally important processes of denitrification, sulfate reduction and acetogenesis, respectively. Non-respiratory anaerobes use fermentation to generate energy and reducing power, secreting metabolic by-products (such as ethanol in brewing) as waste. Facultative anaerobes can switch between fermentation and different terminal electron acceptors depending on the environmental conditions in which they find themselves. As an alternative to heterotrophy many bacteria are autotrophic, fixing carbon dioxide into cell mass. Energy metabolism of bacteria is either based on phototrophy or chemotrophy, i. e. the use of either light or exergonic chemical reactions for fueling life processes. Lithotrophic bacteria use inorganic electron donors for respiration (chemolthotrophs) or biosynthesis and carbon dioxide fixation (photolithotrophs), opposed by organotrophs which need organic compounds as electron donors for biosynthetic reactions (and mostly as well as carbon sources).

Common inorganic electron donors are hydrogen, ammonia (leading to nitrification), iron and several reduced sulfur compounds. In both aerobic phototrophy and chemolithotrophy oxygen is used as a terminal electron acceptor, while under anaerobic conditions inorganic compounds are used instead. Most photolithotrophic and chemolithotrophic organisms are autotrophic, meaning that they obtain cellular carbon by fixation of carbon dioxide, whereas photoorganotrophic and chemoorganotrophic organisms are heterotrophic. In addition to carbon, some organisms also fix nitrogen gas (nitrogen fixation). This environmentally important trait can be found in bacteria of nearly all the metabolic types listed above but is not universal. The distribution of metabolic traits within a group of organisms has traditionally been used to define their taxonomy, although these traits often do not correspond with genetic techniques

### Metastable

A classical system is metastable if it is above its minimum-energy state, but requires an energy input before it can reach a lower-energy state; accordingly, a metastable system can act like a stable system, provided that energy inputs (e.g., thermal fluctuations) remain below some threshold. Systems with strong metastability are commonly described as stable. Quantum mechanical effects can permit metastable states to reach lower energies by tunneling, without an energy input; an associated, broader definition of *metastable* embraces all systems that have a long lifetime (by some standard) in a state above the minimum-energy state.

### Meteorologist

Someone who studies the weather. Often, the person on the evening news who talks about the weather forecast is a meteorologist.

### Meteosat

A geostationary weather satellite launched by the European Space Agency (ESA) and now operated by the company Eumetsat. The most recent version of Meteosat was launched in June 1988. It provides weather imaging of the Earth at both visible light and infrared wavelengths.

### Micellar Nanocontainers

Block copolymer micelles are water- soluble biocompatible nanocontainers with great potential for delivering hydrophobic drugs. An understanding of their cellular distribution is essential to achieving selective delivery of drugs at the subcellular level.

### Micelles

Micelles are small, spherical

structures composed of molecules that attract one another to reduce surface tension. The head of the molecule is hydrophilic, meaning it likes water, while the interior portion is hydrophobic, meaning it avoids water.

## Microarray

Tool for studying how large numbers of genes interact with each other and how a cell's regulatory networks control vast batteries of genes simultaneously. Uses a robot to precisely apply tiny droplets containing functional DNA to glass slides. Researchers then attach fluorescent labels to DNA from the cell they are studying. The labeled probes are allowed to bind to cDNA strands on the slides. The slides are put into a scanning microscope to measure ... how much of a specific DNA fragment is present. What can microarrays reliably provide for medicinal research? Where are the greatest needs for improvement to achieve these goals? Clinical, pharmaceutical, academic, and biotechnology researchers can all agree that sensitivity, high throughput, ease of use, reproducibility, automation, and cost are essential—but how do you optimize the wide range of needs and expectations? Future applications of DNA microarrays in medicine and research are now based on three different molecular analyses.

## Microbiotagraphics

Mapping the microbiotic populations present in the human body.

## Microbubbles

Very small encapsulated gas bubbles (diameters of micrometers) that can be used in diagnostic and therapeutic applications. Upon exposure to sufficiently intense ultrasound, microbubbles will cavitate, rupture, disappear, release gas content, etc. Such characteristics of the microbubbles can be used to enhance diagnostic tests, dissolve blood clots, and deliver drugs or genes for therapy. MeSH 2004

## Microchemistry

The development and use of techniques and equipment to study or perform chemical reactions, with small quantities of materials, frequently less than a milligram or a milliliter.

## Microdevices

Narrower terms: integrated microdevices, nanodevices, SED Single Electron Devices. Related terms microelectronics, microfluidics, micro- TAS.

## Micro-Electromechanical Systems

Micro-Electromechanical Systems (MEMS) also known as microsystems—MST) combine electronics with microscale mechanical devices, resulting in microscopic machinery. Our ability to use synthetic means to fabricate materials, devices and systems on the microscale (microtechnology) is well established. MEMS is a branch of microtechnology that has found numerous commercial applications, for instance accelerometers for airbag deployment in automobiles, ink jet printer heads, color projection displays, chemical sensors and scanning probe microscopy. Furthermore, the silicon semiconductor industry has provided numerous methods for fabricating microscale electronic devices. With a little ingenuity, these same fabrication methods can be used to fabricate devices that convert electrical energy into mechanical energy, and vice-versa.

A promising method for enabling us to control matter on the nanoscale is to build small tools and machines that we can use to build even smaller tools and machines until molecular precision is reached and the full potential of nanotechnology can be realized by Nano-Electromechanical Systems (NEMS). NEMS are orders of magnitude smaller and are one of the goals of nanotechnology. "'Mechanical wear' is not an issue for molecular sized machine parts, not pushed beyond their chemical bond strength. 'Lubricants' are not only unnecessary, but indeed disastrous at this scale—steel balls in a transmission. However, micro and larger machine parts will be utilitized in macro products. Here now are engineering parameters for 'larger' moving diamondoid parts (NTM) Tahir Cagin, Jianwei Che, Michael N. Gardos, Amir Fijany and William A. Goddard, III, "A material for MEMS and NEMS Applications Simulation and Experiments on Friction and Wear of Diamond" NanoTechnology

Magazine June 2000." "MEMS Microelectromechanical systems. Generally used to refer to systems that can respond to a stimulus or create physical forces (sensors and actuators) and that have dimensions on the micrometre scale. They are almost exclusively made using the same lithographic techniques that are used to make silicon chips for computing. Miniature accelerometers are the most successful product in this field and are used to trigger air bags in cars. When such systems can be made with nanoscale dimensions they can be classified as NEMS."

### Microelectronics

Narrower terms: MEMS, nanoelectronics, optoelectronics, SED Single Electron Devices. Related terms: molecular electronics, semiconductors

### Microinjection

The insertion of a substance into a cell through a microelectrode. Typical applications include the injection of drugs, histochemical markers (such as horseradish peroxidase or lucifer yellow) and RNA or DNA in molecular biological studies. To extrude the substances through the very fine electrode tips, either hydrostatic pressure (pressure injection) or electric currents (ionophoresis) is employed. A technique for introducing a solution of DNA, protein, or other soluble material into a cell using a fine microcapillary pipet.

### Micromachines

Mechanical objects that are fabricated in the same general manner as integrated circuits. They are generally considered to be between 100 nanometres to 100 micrometres in size, though that is debatable. The applications of micromachines include accelerometers that detect when a car has hit an object and trigger an airbag. Complex systems of gears and levers are another application. The fabrications of these devices is usually done by one or both of two techniques: surface micromachining and bulk micromachining.

### Microelectromechanical Systems

Microelectromechanical Systems (MEMS) is the technology of the very small, and merges at the nanoscale into "Nanoelectromechanical Systems" (NEMS) and Nanotechnology. In Europe, MEMS are often referred to as *Micro Systems Technology (MST)*. It should not be confused with the hypothetical vision of Molecular nanotechnology or Molecular Electronics. ==of a metre) to a millimetre (thousandth of a metre). At these size scales, a human's intuitive sense of physics do not always hold true.

Due to MEMS' large surface area to volume ratio, surface effects such as electrostatics and wetting dominate volume effects such as inertia or thermal mass. They are fabricated using modified silicon fabrication technology (used to make electronics), molding and plating, wet etching (KOH, TMAH) and dry etching (RIE and DRIE), electro discharge machining (EDM), and other technologies capable of manufacturing very small devices. MEMS sometimes go by the names micromechanics, micro machines, or micro system technology (MST).

Companies with strong MEMS programs come in many sizes. The larger firms specialize in manufacturing high volume inexpensive components or packaged solutions for end markets such as automobiles, biomedical, and electronics. The successful small firms provide value in innovative solutions and absorb the expense of custom fabrication with high sales margins. In addition, both large and small companies work in R&D to explore MEMS technology. Complexity and performance of advanced MEMS based sensors are described by different MEMS sensor generations.

## Micromachinery

Micromachines are mechanical objects that are fabricated in the same general manner as integrated circuits. They are generally considered to be between 100 nanometres to 100 micrometres in size, though that is debatable. The applications of micromachines include accelerometers that detect when a car has hit an object and trigger an airbag. Complex systems of gears and levers are another application.

The fabrications of these devices is usually done by one or both of two techniques: surface micromachining and bulk micromachining. Most micromachines act as transducers; in other words, they are either sensors or actuators.

Sensors convert information from the environment into interpretable electrical signals. One example of a micromachine sensor is a resonant chemical sensor. A lightly damped mechanical object vibrates much more at one frequency than any other, and this frequency is called its resonant frequency. A chemical sensor is coated with a special polymer that attracts certain molecules, such as anthrax, and when those molecules attach to the sensor, its mass increases. The increased mass alters the resonant frequency of the mechanical object, which is detected with circuitry.

Actuators convert electrical signals and energy into motion of some kind. The three most common types of actuators are electrostatic, thermal, and magnetic. Electrostatic actuators use the force of electrostatic energy to move objects. Two mechanical elements, one that is stationary (the stator) and one that is movable (the rotor) have two different voltages applied to them, which creates an electric field. The field competes with a restoring force on the rotor (usually a spring force produced by the bending or stretching of the rotor) to move the rotor. The greater the electric field, the farther the rotor will move. Thermal actuators use the force of thermal expansion to move objects. When a material is heated, it expands and amount depending on material properties. Two objects can be connected in such a way that one object is heated more than the other and expands more, and this imbalance creates motion. The direction of motion depends on the connection between the objects. This is seen in a "heatuator", which is a U-shaped beam with one wide arm and one narrow arm. When a current is passed through the object, heat is created. The narrow arm is heated more than the wide arm due to the fact that they have the same current density. Since the two arms are connected at the top, the stretching hot arm pushes in the direction of the cold arm. Magnetic actuators used fabricated magnetic layers to create forces.

### Micron

$10^{-6}$ Symbol is u. One millionth of a meter, or about 1/25,000 of an inch.

### Microparticles

Applications include calibration of flow cytometers, particle and hematology analyzers, confocal Laser scanning microscopes and zetapotential measuring instruments; flow measurements in gases and liquids like Laser Doppler Anemometry (LDA); Particle Dynamics Analysis (PDA), and Particle Image Velocimetry (PIV); medical diagnostics; separation phases for chromatography; support for immobilized enzymes; spacer in liquid crystal displays (LCD's); peptide synthesis; cell separation; tracers in environmental science; model systems in medicine, biochemistry, colloid chemistry, and aerosol research.

### Microphone

A microphone, sometimes referred to as a mike or mic (pronounced "mike"), is an acoustic to electric transducer that converts sound into an electrical signal. Microphones are used in many applications such as telephones, tape recorders, hearing aids, motion picture production, live and recorded audio engineering, in radio and television broadcasting and in computers for recording voice, VoIP and numerous other computer applications.

### Microplate Reader

Created from the tube spectrophotometer designs of the 1970s to save precious antibody samples. At first clumsy and inaccurate, absorbance microplate readers have evolved to pack unbelievable power and precision, replacing cuvette spectrophotometers for most multisample applications. Continuous improvement is enhancing the classic designs to embrace the world of high-throughput (HT) screening and to allow complete analytical automation. To handle the HT range (more than 10 microplates a day or 1,000 assays), many instruments now allow robotic handling of plates "stacked" in accessory plate handlers.

### Microprocessor

Chip containing the logical elements for performing calculations and carrying out stored instructions.

### Microscopy

Microscopy primer, microscope basics, special techniques, tutorials, virtual microscopy Molecular Expressions, National High Field Magnetic Laboratory, Florida State Univ. US, Olympus America, Inc. 2003 http://micro.magnet.fsu.edu/primer/ Narrower terms: atomic force microscopy AFM, Confocal Scanning Laser Scanning Microscopy CLSM, confocal microscopy, cryoelectron microscopy, electron microscopy, fluorescence microscopy, immunoelectron microscopy, ion microscopy, Laser Fluorescence Microscopy LFM, laser scanning microscopy, Multiphoton Laser Scanning Microscopy MLSM, Magnetic Resonance Force Microscopy MRFM, multiple- photon excitation fluorescence microscopy, Near- field Scanning Optical Microscopy NSOM, Scanning Electron Microscopy SEM, Scanning Transmission Electron Microscopy STEM, Scanning Tunneling Microscopy STM, scanning probe microscopy, Surface Plasmon Resonance microscopy, Total Internal Reflectance Fluorescence Microscopy TIR-FM, Transmission Electron Microscopy TEM, two-photon Laser Fluorescence Microscopy, virtual microscopy

### Microspectrophotometry

Analytical technique for studying substances present at enzyme concentrations in single cells, in situ, by measuring light absorption. Light from a tungsten strip lamp or xenon arc dispersed by a grating monochromator illuminates the optical system of a microscope. The absorbance of light is measured (in nanometers) by comparing the difference between the image of the sample and a reference image. MeSH 1990

### Microspheres

Microspheres are generally defined as small spheres made of any material and sized from about 0.5 μm to 100 μm. Similar, but smaller spheres sized 10 to 500 nm are called *nanospheres*. Ideally, microspheres are completely spherical and homogeneous in size, although less perfect particles are often termed microspheres as well. Depending on the preparation method and material used, microspheres show a typical size distribution which often deviates from the monosized ideal.

## Microstructure

The last decade has seen rapid developments in the fabrication, characterization and conceptual understanding of synthetic microstructures in many different material systems including silicon, III-V and II-VI semiconductors, metals, ceramics and organics. The objective of this journal is to provide a common interdisciplinary platform for the publication of the latest research results on all such "nanostructures" with dimensions in the range of 1-100 nm; the unifying theme here being the dimensions of these artificial structures rather than the material system in which they are fabricated.

## Microsystem

A microscale machine that can sense information from the environment and act accordingly. Outside the U.S., it can also refer to microelectromechanical systems (MEMS).

## Microtubule

A cylindrical biological structure made of tubulin (which is a specific type of protein) dimers, and found in cells. Their role is to mechanically support cellular structure (cell rigidity) and the cell movement. Additionally, biological molecular motors, such as kinezin, use microtubules to move along them while transporting proteins in cells. Tubulin dimers in a microtubule are organized in protofilaments along which they bond relatively strongly. The bonds among protofilaments are weaker. Protofilaments can be thought of as one-dimensional chains of tubulin dimers positioned on a cylinder. In a 13-protofilament microtubule (which occurs in human cells), protofilaments are parallel to a microtubule cylindrical axis. In other types of microtubules, with different number of protofilaments, protofilaments spiral around the microtubule, i.e. they can be thought of as cylindrical geodesics. Microtubules are typical example of bilogical self assembly since tubulin dimers are spontaneously assembled in microtubules in cellular surrounding (cytoplasma). The diameter of a microtubule is about 25 nanometers.

### Microwave

A form of electromagnetic radiation that is beyond the range of the *visible light* spectrum. Microwaves have very high frequencies and wavelengths of 1 mm to 50 cm.

### Miller Indices

A set of 3 integers (4 for hexagonal) that designate crystallographic planes, as determined from reciprocals of fractional axial intercepts.

### Miniaturization

Desirable for many technologies for overall cost reduction (including reduction in the amount of reagents and analytes). Important to remember that building space is often the least available and most expensive component of an overall laboratory budget.

### MM2

A molecular mechanics program developed by Norman Allinger and coworkers; the "MM2 model" is the molecular potential energy function described by the equations, rules, and parameters embodied in that program.

### Mobility (electron, and hole)

the proportionality constant between the carrier drift velocity and applied electric field.

### Modulus of elasticity (E)

The ratio of stress to strain when deformation is totally elastic. Also the Young's modulus.

### Mole

A number of instances of something (typically a molecular species) equaling ~6.022 1023. *Mole* ordinarily means gram-mole; a kilogram-mole is ~6.022 1026.

### Molecular Assembler

A molecular assembler has been defined as a single molecule machine that would assemble individual atoms or molecules according to specific instructions to construct a desired product. Some biological molecules the such as a ribosomes fit this definition since working within cell's environment, it receives instructions from mRNA and then assembles specific sequences of amino acids to construct protein molecules. However, the term "molecular assembler" usually refers to theoretical man-made or synthetic devices. They are thought to be highly desirable since they have been theorized to manufacture products with absolute precision and thus without any pollution. However, others have warned that the such a powerful technology might get out of control and begin to compete with natural forms of life on earth. Since synthetic assemblers have never been constructed a lot of controversy exists as to whether they are possible or simply science fiction. Confusion and controversy has also stemmed from their classification as nanotechnology which is a broadly defined terminology. Nanotechnology is an active area of research which has already been applied to the production of real products; however there are currently no research efforts into the actual construction of "molecular assemblers". A primary criticism of the computational research into "molecular assemblers" is that the structures investigated are thought to be impossible to synthesize.

## Molecular Beam Epitaxy (MBE)

Process used to make compound (multi-layer) semiconductors. Consists of depositing alternating layers of materials, layer by layer, one type after another (such as the semiconductors gallium arsenide and aluminum gallium arsenide).

## Molecular Combing

A method ... which can straighten and align molecules of genomic DNA on a solid surface. The technology also includes a battery of novel statistical methods developed for analyzing the large amounts of data obtained from FISH analyses made on individual DNA molecules. Molecular combing relies on the action of a receding air/ water interface, or meniscus, to uniformly straighten and align DNA molecules on a solid surface. The advantages of this approach reside in the reproducibility of the results, their precision (1 to 4 kb resolution) and the relative ease of analysis afforded by the ability to visualize the molecules directly. Beyond its obvious applications to genomic studies and genetic diseases, it creates new experimental possibilities for research into cancer. Indeed, as a tool, molecular combing is a versatile approach to a wide range of subjects and questions of fundamental interest. This is especially true for the multifaceted domain of DNA replication in eukaryotes.

## Molecular Design

The application of all techniques leading to the discovery of new chemical entities with specific properties required for the intended application.

## Molecular Distillation

A special method for transmission electron microscopy sample preparation. It is especially useful for immunogold labeling. This technology is being developed in the facility.

## Molecular Electronics (ME)

Molecular electronics offers the tantalizing prospect of eventually building circuits with critical dimensions of a few nanometers. Some basic devices utilizing molecules have been demonstrated, including tunnel junctions with negative differential resistance, rectifiers and electrically configurable switches that have been used in simple electronic memory and logic circuits. A major challenge that remains is to show that such devices can be fabricated economically using a process that will scale to circuits with large numbers of elements while maintaining their desired electronic properties.

## Molecular Machine

A mechanical device that performs a useful function using components of nanometer scale and defined molecular structure; includes both artificial nanomachines and naturally occurring devices found in biological systems. Any machine with atomically precise parts of nanometer dimensions; can be used to describe molecular devices found in nature.

## Molecular Manipulator

A device combining a proximal probe mechanism for atomically precise positioning with a molecule binding site on the tip; can serve as the basis for building complex structures by positional synthesis. A programmable device able to position molecular tools with high precision, for example, to direct a sequence of mechanosynthetic steps; a molecular assembler.

## Molecular Manufacturing

Manufacturing using molecular machinery, giving molecule-by-molecule control of products and by-products via positional chemical synthesis. Exponential general-purpose molecular manufacturing — that's a mouthful, but what does it mean? Let's take the phrase apart to see why it is so important. The building of complex structures by mechanochemical processes. The production of complex structures via nonbiological mechanosynthesis.

## Molecular Mechanical Components

In order to lay a foundation for molecular manufacturing, it is necessary to create and to analyze possible designs for nanoscale mechanical components that could, in principle, be manufactured. Because these components could not yet be built in 1998, such designs could not be subjected to rigorous experimental testing and validation. Designers were forced instead to rely upon ab initio structural analysis and computer studies including molecular dynamics simulations. Noted Drexler: "Our ability to model molecular machines (systems and devices) of specific kinds, designed in part for ease of modeling, has far outrun our ability to make them. Design calculations and computational experiments enable the theoretical studies of these devices, independent of the technologies needed to implement them." In nanoscale design, building materials do not change continuously as they are cut and shaped, but rather must be treated as being formed from discrete atoms. A nanoscale component is a supermolecule, not a finely divided solid. Any stray atoms or molecules within such a structure may act as dirt that can clog and disable the device, and the scaling of vibrations, electrical forces, thermal expansion, magnetic interaction and surface tension with size lead to dramatically different phenomena as system size shrinks from the macroscale to the nanoscale. Molecular bearings are perhaps the most convenient class of components to design because their structure and operation is fairly straightforward. One of the simplest examples is Drexler's overlap-repulsion bearing design, shown with end views and exploded views, using both ball-and-stick and space-filling representations. This bearing has exactly 206 atoms including carbon, silicon, oxygen and hydrogen, and is comprised of a small shaft that rotates within a ring sleeve measuring 2.2 nm in diameter. The atoms of the shaft are arranged in a 6-fold symmetry, while the ring has 14-fold symmetry, a combination that provides low energy barriers to shaft rotation.

An exploded view of a 2808-atom strained-shell sleeve bearing designed by Drexler and Merkle using molecular mechanics force fields to ensure that bond lengths, bond angles, van der Waals distances, and strain energies are reasonable. This 4.8-nm diameter bearing features an interlocking-groove interface which derives from a modified diamond {100} surface. Ridges on the shaft interlock with ridges on the sleeve, making a very stiff structure. Attempts to bob the shaft up or down, or rock it from side to side, or displace it in any direction (except axial rotation, wherein displacement is-extremely smooth) encounter a very strong resistance. Whether these bearings would have to be assembled in unitary fashion, or instead could be assembled by inserting one part into the other without damaging either part, had not been extensively studied or modeled by 1998.

Molecular gears are another convenient component system for molecular manufacturing design-ahead. For example, Drexler and Merkle designed a 3557-atom planetary gear, shown in side, end, and exploded views. The entire assembly has twelve moving parts and is 4.3 nm in diameter and 4.4 nm in length, with a molecular weight of 51,009.844 daltons and a molecular volume of 33.458 $nm^3$.

An animation of the computer simulation shows the central shaft rotating rapidly and the peripheral output shaft rotating slowly. The small planetary gears rotate around the central shaft, and they are surrounded by a ring gear that holds the planets in place and ensures that all of the components move in the proper fashion. The ring gear is a strained silicon shell with sulfur atom termination; the sun gear is a structure related to an oxygen-terminated diamond {100} surface; the planet gears resemble multiple hexasterane structures with oxygen rather than $CH_2$ bridges between the parallel rings; and the planet carrier is adapted from a Lomer dislocation array created by R.

## Molecular Mechanics Models

Many of the properties of molecular systems are determined by the molecular potential energy function. Molecular mechanics models approximate this function as a sum of 2-atom, 3-atom, and 4-atom terms, each determined by the geometries and bonds of the component atoms. The 2-atom and 3-atom terms describing bonded interactions roughly correspond to linear springs.

## Molecular Medicine

A variety of pharmaceutical techniques and gene therapies that address specific molecular diseases or molecular defects in biological systems. A variety of pharmaceutical techniques and therapies in use today. Studying molecules as they relate to health and disease, and manipulating those molecules to improve the diagnosis, prevention, and treatment of disease.

## Molecular Mimicry

The process in which structural properties of an introduced molecule imitate or simulate molecules of the host. Direct mimicry of a molecule enables a viral protein to bind directly to a normal substrate as a substitute for the homologous normal ligand. Immunologic molecular mimicry generally refers to what can be described as antigenic mimicry and is defined by the properties of antibodies raised against various facets of epitopes on the viral protein.

## Molecular Motor

Protein or protein complex that transforms chemical energy into mechanical work at a molecular (nanometer) scale. The chemical process that produces energy is hydrolysis of adenosine triphosphate (ATP) to adenosine diphosphate (ADP) and phosphate (P). Rotatory and translatory motors are known to exist.

Motor proteins move in eukaryotic cells along filaments (actin filaments, microtubules)—periodic and relatively rigid protein structures with a periodicity of the order of 10 nanometers. Some of the known motors are myosin that moves along actin filaments, and kinesin and dynein that move along microtubules (tubulin filaments). They play a role in muscular contraction, cell division, cellular material transport, cell division, etc. The fact that such structures exist is a great inspiration for nanotechnological applications (including nanobots, of course).

## Molecular Nanoscience

An emerging interdisciplinary field that combines the study of molecular/ biomolecular systems with the science and technology of nanoscale structures and systems. The potential applications for this research are very broad and include such possibilities as the use of biomolecules and cellular systems to self- assemble nanoelectronic circuitry and other nanoscale structures and the use of lamellar host frameworks containing nano pores that can be tailored to include guest molecules for separation of chemicals for pharmaceutical and other applications.

## Molecular Nanotechnology (MNT)

Molecular nanotechnology (MNT) is the engineering of functional systems at the molecular scale. An equivalent definition would be "machines at the molecular scale designed and built atom-by-atom". This is distinct from nanoscale materials. Based on Richard Feynman's vision of miniature factories using nanomachines to build complex products (including additional nanomachines), this advanced form of nanotechnology (or *molecular manufacturing*) will make use of positionally-controlled mechanosynthesis guided by molecular machine systems. MNT would involve combining physical principles demonstrated by chemistry, other nanotechnologies, and the molecular machinery of life with the systems engineering principles found in modern macroscale factories. Its most well-known exposition is in the books of K. Eric Drexler.

Formulating a roadmap for the development of MNT is now an objective of a broadly based technology roadmap project led by Battelle (the manager of several U.S. National Laboratories) and the Foresight Institute. The roadmap should be completed by early 2007. In August 2005, a task force consisting of 50+ international experts from various fields was organized by the Center for Responsible Nanotechnology to study the societal implications of molecular nanotechnology.

While conventional chemistry employs stochastic processes driven toward some equilibrium to obtain stochastic results, and biology exploits stochastic processes to obtain deterministic results based on complex enzyme-catalyzed reaction chains optimized through billions of years of evolutionary feedback, molecular nanotechnology would employ novel (and as yet unspecified) deterministic nanoscale processes to obtain deterministic results. The desire in molecular nano-technology would be to place molecular moieties in deterministic locations with deterministic orientation to obtain desired chemical reactions, and then to build systems by further assembling the products of these reactions.

Ralph Merkle has compared today's manufacturing methods (in contrast to mechano-synthesis) to an attempt to build interesting Lego brick constructions while wearing boxing gloves: "Casting, grinding, milling and even lithography move atoms in great thundering statistical herds. It's like trying to make things out of LEGO blocks with boxing gloves on your hands. Yes, you can push the LEGO blocks into great heaps and pile them up, but you can't really snap them together the way you'd like." It has been posited that molecular nanotechnology could offer much cleaner manufacturing processes than today's bulk technology.

One proposed application of MNT is the development of so-called smart materials. This term refers to any sort of material designed and engineered at the nanometer scale to perform a specific task, and encompasses a wide variety of possible commercial applications. One example is materials designed to respond differently to various molecules; such a capability could lead, for example, to artificial drugs which would recognize and render inert specific viruses. Another is the idea of self-healing structures, which would repair small tears in a surface naturally in the same way as self-sealing tires or human skin; and while this technology is relatively new, it is already seeing commercial application in various engineering plastics.

A nanosensor created by MNT would resemble a smart material, involving a small component within a larger machine that would react to its environment and change in some fundamental, intentional way. As a very simple example: a photosensor could passively measure the incident light and discharge its absorbed energy as electricity when the light passes above or below a specified threshold, sending a signal to a larger machine. Such a sensor would cost less and use less power than a conventional sensor, and yet function usefully in all the same applications — for example, turning on parking lot lights when it gets dark.

While smart materials and nanosensors both exemplify useful applications of MNT, they pale in comparison with the complexity of the technology most popularly associated with the term: the replicating nanorobot.

### *Medical Nanorobots*

One of the most important applications of MNT would be medical nanorobotics or nanomedicine, an area pioneered by Robert Freitas in numerous books and papers.

The ability to design, build, and deploy large numbers of medical nanorobots would, at an optimum, make possible the rapid elimination of disease and the reliable and relatively painless recovery from physical trauma. Medical nanorobots might also make possible the convenient correction of genetic defects, and help to ensure a greatly expanded healthspan.

More controversially, medical nanorobots might be used to augment natural human capabilities. However, mechanical medical nanodevices would not be allowed (or designed) to self-replicate inside the human body, nor would medical nanorobots have any need for self-replication themselves since they would be manufactured exclusively in carefully regulated nanofactories.

### *Utility Fog*

Another proposed application of nanotechnology involves utility fog— in which a cloud of networked microscopic robots (simpler than assemblers) changes its shape and properties to form macroscopic objects and tools in accordance with software commands. Rather than modify the current practices of consuming material goods in different forms, utility fog would simply replace most physical objects.

### *Phased-Array Optics*

Yet another proposed application would be phased-array optics (PAO). PAO would used the principle of phased-array millimeter technology but at optical wavelengths. This would permit the duplication of any sort of optical effect but virtually. Users could request holograms, sunrises and sunsets, or floating lasers as the mood strikes. PAO systems were described in BC Crandall's *Nanotechnology: Molecular Speculations on Global Abundance* in the Brian Wowk article "Phased-Array Optics".

### *Potential Social Impacts*

Despite the current early developmental status of nanotechnology and molecular nanotechnology, much concern surrounds MNT's anticipated impact on economics and on law. Some conjecture that MNT would elicit a strong public-opinion backlash, as has occurred recently around genetically modified plants and the prospect of human cloning. Whatever the exact effects, MNT, if achieved, would tend to upset existing economic structures by reducing the scarcity of manufactured goods and making many more goods (such as food and health aids) manufacturable.

It is generally considered that future citizens of a molecular-nanotechnological society would still need money, in the form of unforgeable digital cash or physical specie (in special circumstances). They might use such money to buy goods and services that are unique, or limited within the solar system. These might include: matter, energy, information, real estate, design services, entertainment services, legal services, fame, political power, or the attention of other people to your political/religious/philosophical message. Furthermore, futurists must consider war, even between prosperous states, and non-economic goals.

If MNT were realized, some resources would remain limited, because unique physical objects are limited (a plot of land in the real Jerusalem, mining rights to the larger near-earth asteroids) or because they depend on the goodwill of a particular person (the love of a famous person, a painting from a famous artist). Demand will always exceed supply for some things, and a political economy may continue to exist in any case. Whether the interest in these limited resources would diminish with the advent of virtual reality, where they could be easily substituted, is yet unclear; one reason why it might not is hypothetical irrational preference for "the real thing".

*Replicating Nanobots*

MNT nanofacturing is popularly linked with the idea of swarms of coordinated nanoscale robots working together, as proposed by Drexler in his 1986 popular discussions of the subject. It is proposed that sufficiently capable nanobots could construct more nanobots.

However, critics doubt both the feasibility of self-replicating nanobots and the feasibilty of control if self-replicating nanobots could be achieved: they cite the possibility of mutations removing any control and favoring reproduction of mutant pathogenic variations. Advocates address the second doubt by arguing that bacteria are (of necessity) evolved to evolve, while nanobot mutation can be actively prevented by common error-correcting techniques. Similar ideas are advocated in the Foresight Guidelines on Molecular Nanotechnology.

Recent technical proposals for MNT nanofactories do not include self-replicating nanobots, and recent ethical guidelines prohibit self-replication.

*Risks*

Molecular nanotechnology is one of the technologies that some analysts believe could lead to a Technological Singularity.

Aside from the fantasy scenarios, some feel that molecular nanotechnology would have daunting risks. It conceivably could enable cheaper and more destructive conventional weapons. Also, molecular nanotechnology might permit weapons of mass destruction that could self-replicate, as viruses and cancer cells do when attacking the human body. Commentators generally agree that, in the event molecular nanotechnology were developed, humankind should permit self-replication only under very controlled or "inherently safe" conditions.

A fear exists that nanomechanical robots, if achieved, and if designed to self-replicate using naturally occurring materials (a difficult task), could consume the entire planet in their hunger for raw materials, or simply crowd out natural life, out-competing it for energy (as happened historically when blue-green algae appeared and outcompeted earlier life forms). Some commentators have referred to this situation as the "grey goo" or "ecophagy" scenario. K. Eric Drexler considers an accidental "grey goo" scenario extremely unlikely and says so in later editions of *Engines of Creation*. The "grey goo" scenario begs the Tree Sap Answer: what chances exist that one's car could spontaneously mutate into a wild car, run off-road and live in the forest off tree sap? In light of this perception of potential danger, the Foresight Institute (founded by K. Eric Drexler to prepare for the arrival of future technologies) has drafted a set of guidelines for the ethical development of nanotechnology. These include the banning of free-foraging self-replicating pseudo-organisms on the Earth's surface, at least, and possibly in other places.

*Technical Issues and Criticism*

Universal Assemblers vs. Diamondoid Nanofactories A section heading in Drexler's *Engines of Creation* reads "Universal Assemblers", and the following text speaks of molecular assemblers which could hypothetically "build almost anything that the laws of nature allow to exist." Drexler's colleague Ralph Merkle has noted that, contrary to widespread legend, Drexler never claimed that assembler systems could build absolutely any molecular structure. The endnotes in Drexler's book explain the qualification "almost": "For example, a delicate structure might be designed that, like a stone arch, would self-destruct unless all its pieces were already in place. If there were no room in the design for the placement and removal of a scaffolding, then the structure might be impossible to build. Few structures of practical interest seem likely to exhibit such a problem, however."

In 1992, Drexler published *Nanosystems: molecular machinery, manufacturing, and computation*, a detailed proposal for synthesizing stiff, diamond-based structures using a table-top factory. Although such a nanofactory would be far less powerful than a protean universal assembler, it would still be enormously capable. Diamondoid structures and other stiff covalent structures, if achieved, would have a wide range of possible applications, going far beyond current MEMS technology. However, no proposal was put forward for building the table-top factory in the absence of a near-universal assembler.

*The Smalley-Drexler Debate*

Several researchers, including Dr. Smalley, have attacked the notion of universal assemblers, leading to a rebuttal from Drexler and colleagues, and eventually to an exchange of letters. Smalley argues that chemistry is extremely complicated, reactions are hard to control, and that a universal assembler is science fiction. Drexler and colleagues, however, note that Drexler never proposed universal assemblers able to make absolutely anything, but had instead proposed more limited assemblers able to make a very wide variety of things. They challenge the relevance of Smalley's arguments to the more specific proposals advanced in *Nanosystems*.

The feasibility of Drexler's proposals largely depends, therefore, on whether designs like those in *Nanosystems* could be built in the absence of a universal assembler to build them and would work as described. Supporters of molecular nanotechnology frequently claim that no significant errors have been discovered in *Nanosystems* since 1992. Even some critics concede that "Drexler has carefully considered a number of physical principles underlying the 'high level' aspects of the nanosystems he proposes and, indeed, has thought in some detail" about some issues.

Other critics claim, however, that *Nanosystems* omits important chemical details about the low-level 'machine language' of molecular nanotechnology (Smalley, Atkinson, Moriarty, Jones). They also claim that much of the other low-level chemistry in *Nanosystems* requires extensive further work, and that Drexler's higher-level designs therefore rest on speculative foundations.

Drexler argues that we may need to wait until our conventional nanotechnology improves before solving these issues: "Molecular manufacturing will result from a series of advances in molecular machine systems, much as the first Moon landing resulted from a series of advances in liquid-fuel rocket systems. We are now in a position like that of the British Interplanetary Society of the 1930s which described how multistage liquid-fuelled rockets could reach the Moon and pointed to early rockets as illustrations of the basic principle."

In any case, as Richard Feynman once said, "It is scientific only to say what's more likely or less likely, and not to be proving all the time what's possible or impossible."

The fundamental question to address is whether or not most structures consistent with physical law can in fact be manufactured. Note that this framing of the question avoids endless digressions about what Drexler did or did not propose, which are ultimately beside the point. Either (a) such a manufacturing capabiliity is feasible, in which case the focus should be on what it looks like and how to develop it, or (b) such a manufacturing capability is for some reason infeasible, in which case the focus should be on understanding why it can not be done.

The most recent paper in this continuing research effort by Freitas, Merkle and their collaborators reports that the most-studied mechanosynthesis tooltip motif (DCB6Ge) successfully places a $C_2$ carbon dimer on a C(110) diamond surface at both 300K (room temperature) and 80K (liquid nitrogen temperature), and that the silicon variant (DCB6Si) also works at 80K but not at 300K. These tooltips are intended to be used only in carefully controlled environments (e.g., vacuum). Maximum acceptable limits for tooltip translational and rotational misplacement errors are reported in paper III — tooltips must be positioned with great accuracy to avoid bonding the dimer incorrectly. Over 100,000 CPU hours were invested in this latest study. The DCB6 tooltip motif, initially described at a Foresight Conference in 2002, was the first complete tooltip ever proposed for diamond mechanosynthesis and remains the only tooltip motif that has been successfully simulated for its intended function on a full 200-atom diamond surface.

Further research to consider additional tooltips will require time-consuming computational chemistry and difficult laboratory work.

A working nanofactory would require a variety of well-designed tips for different reactions, and detailed analyses of placing atoms on more complicated surfaces. Although this appears a challenging problem given current resources, many tools will be available to help future researchers: Moore's Law predicts further increases in computer power, semiconductor fabrication techniques continue to approach the nanoscale, and researchers grow ever more skilled at using proteins, ribosomes and DNA to perform novel chemistry.

## Molecular Recognition

The ability of biological macromolecules such as proteins and DNA to recognize selectively and to bind to other species to form larger supramolecular complexes is a key element in the extraordinarily diverse and controlled chemistry exhibited by nature. Chemists today are increasingly interested in mimicking these processes which involve 'molecular recognition' of one molecule by another via formation of specific nonequivalent bonds between them and spontaneous binding together of two to many thousands of molecules into well defined supramolecular systems with new chemical properties. Industrially important applications already include drug design, synthesis of stereo regular polymers, affinity chromatography, crystallization, epitaxial growth and liquid crystal displays.

## Molecular Self-assembly

The spontaneous formation of molecules into covalently bonded, well- defined, stable structures—is a very important concept in biological systems and has increasingly become a focus of non- biological research.

## Molecular Sorting Rotor

A class of nanomechanical device capable of selectively binding (or releasing) molecules from (or to) solution, and of transporting these bound molecules against significant concentration gradients.

## Molecular Surgery (Molecular Repair)

In medical nanorobotics, the analysis and physical correction of molecular structures in the body using medical nanomachines. Analysis and physical correction of molecular structures in the body using medical nanomachines.

## Molecular Systems Engineering

Design, analysis, and construction of systems of molecular parts working together to carry out a useful purpose.

### Molecular Wire

A molecular wire—the simplest electronic component—is a quasi-one-dimensional molecule that can transport charge carriers (electrons or holes) between its ends.

### Molecularly Imprinted Polymers MIPs

A new class of materials that have artificially created receptor structures. Since their discovery in 1972, MIPs have attracted considerable interest from scientists and 3engineers involved with the development of chromatographic absorbents, membranes, sensors and enzyme and receptor mimics.

### Molecule

A set of atoms linked by covalent bonds. A macroscopic piece of diamond is technically a single molecule. (Sets of atoms linked by bonds of other kinds are sometimes also termed molecules.) Group of atoms held together by chemical bonds; the typical unit manipulated by nanotechnology. The smallest particle of a chemical substance; typically a group of atoms held together in a particular patter, by chemical bonds. Molecules are groups of atoms which are bonded together. How molecules are formed is governed by the laws of physics. An illustration representing a molecule: above a gold surface. The grey balls are: carbon atoms, the little yellow balls are sulfur,: the big yellow balls are the gold,: the white are hydrogen, and the rods joining: the balls together represent the bonds.

### Moledular Technology

Moledular technology promises to create a new class of imaging agents that offer distinct advantages and can be used for the early detection and diagnosis of disease. Diagnostic molecules have the potential to act as biomarkers in drug development and diagnostics, and can be used in the imaging of cancer in living subjects. This meeting will address the challenges in implementing nanotechnology for drug delivery systems and imaging agents, and promote dialogue between diagnostic and therapeutic development. The production and application of structures, devices and systems by controlling shape and size at nanometre scale. The development and use of techniques to study physical phenomena and construct structures in the nanoscale size range or smaller. The creation of functional materials, devices and systems through control of matter at the scale of 1 to 100 nanometers, and the exploitation of novel properties and phenomena at the same scale. Nanotechnology is emerging as a field critical for enabling essential breakthroughs that may have tremendous potential for affecting biomedicine. Moreover, nanotechnologies developed in the next several years may well form the foundation of significant commercial platforms. Emerging as a new field enabling the creation and application of materials, devices, and systems at atomic and molecular levels and the exploitation of novel properties that emerge at the nanometer scale. Many areas of biomedicine are expected to benefit from nanotechnology including sensors for use in the laboratory, the clinic, and within the human body; new formulations and routes for drug delivery; and biocompatible, high- performance materials for use in implants.

Examples of potential uses of nanotechnology in biomedicine include the early detection and treatment of disease and the development of "smart", rejection- resistant implants that will respond appropriately as the body's needs change. Although research in this field dates back to Richard P. Feynman's classic talk in 1959, the term nanotechnology was first coined by K. Eric Drexler in 1986 in the book *Engines of Creation*. In the popular press, the term nanotechnology is sometimes used to refer to any sub- micron process, including lithography. Because of this, many scientists are beginning to use the term *molecular nanotechnology* when talking about true nanotechnology at the molecular level.

## Moledulcar Manufacturing

The automated building of products from the bottom up, molecule by molecule, with atomic precision. This will make products that are extremely lightweight, flexible, durable, and potentially very 'smart'.

## Monkeywrenching

In medical nanorobotics, the mechanical or chemical jamming of cellular equilibrium processes, with a cytocidal objective.

## Monoclonal Antibodies

Produced by injecting animals to elicit a response from lymphocytes to produce antibodies. Lymphocytes which produce antibodies with strong binding capability can be isolated and used to produce only one kind of antibody (monoclonal) on a permanent basis once the lymphocytes are immortalized. This is accomplished by fusing them (combining them genetically) with cancer cells which have the distinction of living indefinitely in a culture. Monoclonal antibodies can be produced repeatedly and collected for use in immunodetection.

## Monolayer

A monolayer is a single, closely packed layer of atoms or molecules. A Langmuir monolayer or unsoluble monolater is a one-molecule thick insoluble layer of an organic material spread onto an aqueous subphase. Traditional compounds used to prepare Langmuir monolayers are amphiphilic materials that possess a hydrophilic headgroup and a hydrophobic tail. Since the 1980s a large number of other materials have been employed to produce Langmuir monolayers, some of which are semi-amphiphilic, including macromolecules such as polymers. Langmuir monolayers are extensively studied for the fabrication of Langmuir-Blodgett film (LB films), which are formed by transferred monolayers on a solid substrate.

A Gibbs monolayer or soluble monolayer is a monolayer formed by a compound that is soluble in one of the phases separated by the interface on which the monolayer is formed.

In biology monolayers are frequently encountered in for instance the phospholipid lipid bilayer structure of biological membranes.

## Monomer

The units from which a polymer is constructed.

## Monomolecular Computing

The implantation inside a single molecule of ALL the functional groups or circuits to realize a calculation, without any help from external artifices such as re-configuration, calculation sharing between the user and the machine, or selection of the operational devices.

## Most-probable Loss (MPL) Tomography

A novel method for acquiring data for electron tomography dramatically increases resolution and usable section thickness of stained samples, according to a NCMIR study published in the December 2004 issue of the *Journal of Structural Biology*. ... takes advantage of a new generation of energy-filtering intermediate voltage electron microscopes (IVEMs), such as NCMIR's JEM-3200EF IVEM (300 kV).

## MSat

A powerful communications satellite launched by the Canadian firm TMI Communications on April 20, 1996. MSat was Canada's first satellite designed to serve mobile users, especially those in remote areas out of the reach of conventional communication systems.

## Multiphoton Fluorescence Microscopy

Flurorescence microscopy utilizing multiple low- energy photons to produce the excitation event of the fluorophore. Multiphoton microscopes have a simplified optical path in the emission side due to the lack of an emission pinhole, which is necessary with normal confocal microscopes.

Ultimately this allows spatial isolation of the excitation event, enabling deeper imaging into optically thick tissue, while restricting photobleaching and photoxicity to the area being imaged.

## Multiple-photon Excitation Fluorescence Microscopy

That is a a technique uses non- linear optical effects to achieve optical sectioning. ... Advantages of multiphoton imaging: Optical sections may be obtained from deeper within a tissue that can be achieved by confocal or wide- field imaging. There are three main reasons for this: the excitation source is not attenuated by absorption by fluorophore above the plane of focus longer excitation wavelengths suffer less scattering fluorescence signal is not degraded by scattering from within the sample as it is not imaged.

# N

### Nano Cubic Technology

An ultra-thin layer coating that results in higher resolution for recording digital data, ultra-low noise and high signal-to-noise ratios that are ideal for magneto-resistive (MR) heads. It is capable of catapulting data cartridge and digital videotape to one-terabyte native (uncompressed) capacities and floppy disk capacities to three gigabytes. To help visualize the potential, 1TB can store up to 200 two-hour movies.

### Nanoarray

An ultra-sensitve, ultra-miniaturized array for biomolecular analysis. BioForce Nanosciences' Nanoarrays utilize approximately 1/10,000$^{th}$ of the surface area occupied by a conventional microarray, and over 1500 nanoarray spots can be placed in the area occupied by a single microarray domain.

### Nanoassembler

The Holy Grail of nanotechnology; once a perfected nanoassembler is availble, building anything becomes possible, with physics and the imagination the only limitation (of course each item would have to be designed first, which is another small hurdle).

### Nanobiology

Fundamental biological functions are carried out by molecular machineries that have the sizes of 1-100 nm. You find many examples in molecular biology and cell biology: single enzymes, transcription complex, ribosome, transport complex, nuclear pore, and so on.

To understand the functions of these machineries, one has to describe their movements, changes in their shapes, and their localization.

This means the mechanistic study is equivalent to dynamic morphology at this level of size, making a new field that merges mechanistic biology and morphology.

## Nanobioprocessors

Implantable nano scale processors that can integrate with biological pathways and modify biological processes. Trans- NIH Bioengineering Nano-technology Initiative, SBIR, PA Number 02- 125.

## Nanobiotechnology

Nanobiotechnology is the branch of nanotechnology with biological and biochemical applications or uses. Examples include nanosensors based on biomolecules such as DNA or proteins.

## Nanobubbles

Tiny air bubbles on colloid surfaces. Thought to reduce drag, such as would be of benefit to swimmers wearing a suit coverd in them.

## Nanocantilever

The simplest micro-electromechanical system (MEMS) that can be easily machined and mass-produced via the same techniques used to make computer chips. The ability to detect extremely small displacements make nanocantilever beams an ideal device for detecting extremely small forces, stresses and masses. Nanocantilevers coated with antibodies, for example, will bend from the mass added when substrate binds to its antibody, providing a detector capable of sensing the presence of single molecules of clinical importance.

## Nanocentrifuge

In medical nanorobotics, a proposed nanodevice that can spin materials at very high speed, imparting rotational accelerations of up to one trillion gravities (g's), thus permitting rapid sortation.

## Nanochemistry

The goal of the Nanochemistry group is to develop chemistry beyond the realm of individual molecules by exploring how systems of several molecules or other electronically confined entities self-organize into distinct structures on the nanometer scale. The new insight is used to study and manipulate biological nanosystems and to create novel nanoarchitectures with unique structural, electronic, optical and magnetic properties. Specific research areas include bionanotechnology and self- assembling molecular electronics.

## Nanochip

Nanochip is an integrated circuit ( IC ) that is so small, in physical terms, that individual particles of matter play major roles. Miniaturization of electronic and computer components has always been a primary goal.

## Nanochondria

Nanomachines existing inside living cells, participating in their biochemistry and/or assembling various structures.

## Nanochronometer

In medical nanorobotics, a proposed clock or timing mechanism constructed of nanoscale components.

## Nanocircles

A nanocluster or nanocrystal is a fragment of solid comprising somewhere between a few atoms to a few tens of thousands of atoms. Nanoclusters are therefore a novel state of matter with properties that are neither those of a bulk crystal nor those of individual atoms and molecules. Over the past 10 years huge advances have been made both in the synthesis of size- tunable, monodisperse (i.e. size- selected) nanoclusters of various chemical compositions and in the development of techniques for their assembly into nanostructured solids (facilitating the synthesis of what have been termed "designer materials").

## Nanoclusters

A nanocluster or *nanocrystal* is a fragment of solid comprising somewhere between a few atoms to a few tens of thousands of atoms. Nanoclusters are therefore a novel state of matter with properties that are neither those of a bulk crystal nor those of individual atoms and molecules. Over the past 10 years huge advances have been made both in the synthesis of size- tunable, monodisperse (i.e. size- selected) nanoclusters of various chemical compositions and in the development of techniques for their assembly into nanostructured solids (facilitating the synthesis of what have been termed "designer materials").

### Nanocones

Nonplanar graphitic structures. Carbon-based structures with five-fold symmetry that form due to disclination defects in two-dimensional graphene sheets. They have been observed as nanotube caps and as freestanding structures.

### Nanocontainers

"Micellar nanocontainers" or "Micelles," these are nanoscale polymeric containers that could be used to selectively deliver hydrophobic drugs to specific sites within individual cells.

### Nanocrystals

A nanocrystal is a crystalline material with dimensions measured in nanometers; a nanoparticle with a structure that is mostly crystalline. These materials are of huge technological interest since many of their electrical and thermodynamic properties show strong size dependence, and can therefore be controlled through careful manufacturing processes. Nanocrystals are also of interest since they often provide single domain crystalline systems that can be studied to provide information that can help explain the behaviour of macroscopic samples of similar materials, without the complicating presence of grain boundaries and other defects. Semiconductor nanocrystals in the sub 10nm size range are often referred to as quantum dots.

Nanocrystals made with zeolite are used as a filter to turn crude oil onto diesel fuel at an ExxonMobil oil refinery in Louisiana, a method cheaper than the conventional way. A layer of nanocrystals is applied to new type of solar panel named SolarPly made by Nanosolar. It is cheaper than other solar panels, more flexible, and claims 12% efficiency. (Conventional solar panels convert 9% of the sun's energy into electricity.) Crystal tetrapods 40 nanometers wide convert photons into electricity, but only have 3% efficiency.

### Nanocrystalline Silicon

Nanocrystalline silicon (nc-Si) - an allotropic form of silicon - is similar to amorphous silicon (a-Si), in that it has an amorphous phase. Where they differ, however, is that nc-Si has small grains of crystalline silicon within the amorphous phase. This is in contrast to polycrystalline silicon (poly-Si) which consists solely of crystalline silicon grains, separated by grain boundaries. nc-Si is sometimes also known as microcrystalline silicon (μc-Si). The difference comes solely from the grain size of the crystalline grains. Most materials with grains in the micrometre range are actually fine-grained polysilicon, so nanocrystalline silicon is a better term. nc-Si has many useful advantages over a-Si, one being that if grown properly it can have a higher mobility, due to the presence of the silicon crystallites.

It also shows increased absorption in the red and infrared wavelengths, which make it an important material for use in a-Si solar cells. One of the most important advantages of nanocrystalline silicon, however, is that it has increased stability over a-Si, one of the reasons being because of its lower hydrogen concentration. Although it currently cannot attain the mobility that poly-Si can, it has the advantage over poly-Si that it is easier to fabricate, as it can be deposited using conventional low temperature a-Si deposition techniques, such as PECVD, as opposed to laser annealing or high temperature CVD processes, in the case of poly-Si.

## Nanodevices

Scientists believe nanoscale devices may lead to computer chips with billions of transistors, instead of millions—which is the typical range in today's semiconductor technology. The more transistors crammed on a chip, the more powerful it is. "This technology has the potential to replace existing manufacturing methods for integrated circuits, which may reach their practical limits within the next decade when Moore's Law eventually hits a brick wall," said physicist Bernard Yurke of Bell Labs. DNA, which provides the molecular blueprints for all living cells, is an ideal tool for making nanoscale devices. "We took advantage of how pieces of DNA—with its billions of possible variations—lock together in only one particular way, like pieces of a jigsaw puzzle," Yurke said.

## Nanodots

Nanodots, also known as quantum dots, consist of 100s-1000s of atoms of inorganic semiconductor nanoparticles and are approximately one billionth of a meter in size. Developed in the mid-1980s for optoelectronic applications, they have interesting structural, electronic, and optical properties—they strongly absorb light in the near UV range and re-emit visible light that has its color determined by both the nanodot size and surface chemistry. And as the size of nanodots can be controlled during synthesis with nanoscale precision, so the optical properties can be manipulated. In addition, nanodots have a longer life than organic fluorophores, and have a broad excitation spectrum. These factors combined make the use of quantum dots as light-emitting phosphors a strong candidate for a major application of nanotechnology in the future.

## Nanoelectronics

Nanoelectronics refer to the use of nanotechnology on electronic components, especially transistors. Although the term *nanotechnology* is generally defined as *utilizing technology less than 100nm in size,* nanoelectronics often refer to transistor devices that are so small that inter-atomic interactions and quantum mechanical properties need to be studied extensively. As a result, present transistors (such as CMOS90 from TSMC or Pentium 4 Processors from Intel) do not fall under this category, even though these devices are manufactured under 90nm or 65nm technology[*dubious – discuss*].

Nanoelectronics are sometimes considered as disruptive technology because present candidates are significantly different from traditional transistors. Some of these candidates include: hybrid molecular/semiconductor electronics, one dimensional nanotubes/nanowires, or advanced molecular electronics. The sub-voltage and deep-sub-voltage nanoelectronics are specific and important fields of R&D, and the appearance of new ICs operating almost near theoretical limitTemplate: Cm (fundamental, technological, design methodological, architectural, algorithmic) on energy consumption per 1 bit processing is inevitable. The important case of fundamental ultimate limit for logic operation is reversible computing.

Although all of these hold immense promises for the future, they are still under development and will most likely not be used for manufacturing any time soon.

## Nanoengineering

We use chemistry to construct nanostructures and their composites, then focus our attention on the electronic, optical, and transport properties of these nanostructures and the macroscopic films and materials that can be constructed from them. This research lies at the common frontier of chemistry, condensed matter physics, optics, and bioengineering.

## Nanofactory

A nanofactory is a proposed system in which nanomachines (resembling molecular assemblers, or industrial robot arms) would combine reactive molecules via mechanosynthesis to build larger atomically precise parts. These, in turn, would be assembled by positioning mechanisms of assorted sizes to build macroscopic (visible) but still atomically-precise products.

A typical nanofactory would fit in a desktop box, in the vision of K. Eric Drexler published in *Nanosystems: Molecular Machinery, Manufacturing and Computation* (1992), a notable work of "exploratory engineering". During the last decade, others have extended the nanofactory concept, including an analysis of nanofactory convergent assembly by Ralph Merkle, a systems design of a replicating nanofactory architecture by J. Storrs Hall, Forrest Bishop's "Universal Assembler", the patented exponential assembly process by Zyvex, and a top-level systems design for a 'primitive nanofactory' by Chris Phoenix (Director of Research at the Center for Responsible Nanotechnology). All of these nanofactory designs (and more) are summarized in Chapter 4 of *Kinematic Self-Replicating Machines* (2004) by Robert Freitas and Ralph Merkle.

In 2005, a computer-animated short film of the nanofactory concept was produced by John Burch, in collaboration with Drexler. Such visions have been the subject of much debate, on several intellectual levels. No one has discovered an insurmountable problem with the underlying theories and no one has proved that the theories can be translated into practice. However, the debate continues, with some of it being summarized in the Molecular nanotechnology article. If nanofactories could be built, severe disruption to the world economy would be one of many possible negative impacts. Great benefits also would be anticipated. Various works of science fiction have explored these and similar concepts. The potential for such devices was part of the mandate of a major UK study led by mechanical engineering professor Ann Dowling. The report is complete.

## Nanofibers

Using a proprietary process, eSpin is able to produce minute fibers which are 10 to 100 times smaller in diameter than what is possible with conventional textile technology. For comparison, eSpin nanofibers are about 100 times smaller than a human hair. These nanofibers provide a very large surface area for a given weight of fibers. High surface area is at the heart of many envisioned products. Used for clean room products, nanocomposites, filtration, surgical gowns, biomedical devices, and specialty fabrics.

## Nanofluidics

Controlling nano-scale amounts of fluids Researchers at Cornell University are using nanotechnology to build microscopic silicon devices with features comparable in size to DNA, proteins and other biological molecules — to count molecules, analyze them, separate them, perhaps even work with them one at a time.

## Nano-genomics

The main objective of the group is to develop experimental tools that will accelerate the finding of genomic factors (e.g. genes, polymorphisms, methylation patterns) that contribute to complex traits. In the long term this may lead to predictive medicine based on personalized genomic information. Currently available molecular tools are too slow, cumbersome and expensive to allow comprehensive genomic information to be easily obtained from a large number of individuals in a population (e.g. for case/control studies). A radical reinvention of the way genes and DNA are analyzed is needed. We believe that the analysis of single DNA molecules can overcome many of the present limitations and in addition will reveal information which is inaccessible by conventional methods.

## Nanoguitar

Smallest guitar, about the size of a human blood cell, carved out of crystalline silicon, illustrates new technology for nanosized electromechanical devices. The world's smallest guitar is 10 micrometers long — about the size of a single cell — with six strings each about 50 nanometers, or 100 atoms, wide. Just one of several structures that Cornell researchers believe are the world's smallest silicon mechanical devices. Researchers made these devices at the Cornell Nanofabrication Facility, bringing microelectromechanical devices, or MEMS, to a new, even smaller scale — the nano-sized world."

## Nanoharvesting

Nanoharvesting agents are being designed to act as molecular mops for these cancer biomarkers and can be directly queried by mass spectrometry.

## Nanohorns

One of the single walled carbon nanotube types, with an irregular horn-like shape, which may be a critical component of a new generation of fuel cells. "The main characteristic of the carbon nanohorns is that when many of the nanohorns group together an aggregate (a secondary particle) of about 100 nanometers is created. The advantage being, that when used as an electrode for a fuel cell, not only is the surface area extremely large, but also, it is easy for the gas and liquid to permeate to the inside. In addition, compared with normal nanotubes, because the nanohorns are easily prepared with high purity it is expected to become a low-cost raw material."

## Nanoimprint Lithography

Nanoimprint lithography is a novel method of fabricating nanometer scale patterns. It is a simple process with low cost, high throughput and high resolution. It creates patterns by mechanical deformation of imprint resist and subsequent processes. The imprint resist is typically a monomer or polymer formulation that is cured by heat or UV light during the imprinting. Adhesion between the resist and the template is controlled to allow proper release.

Nanoimprint lithography was first invented by Prof. Stephen Chou and his students. Soon after its invention, a lot of researchers developed many different variations and implementations. At this point, nanoimprint lithography has been added to the International Technology Roadmap for Semiconductors (ITRS) for the 32 nm node.

Nanoimprint lithography was first invented by Prof. Stephen Chou and his students. Soon after its invention, a lot of researchers developed many different variations and implementations. At this point, nanoimprint lithography has been added to the International Technology Roadmap for Semiconductors (ITRS) for the 32 nm node.

Thermoplastic Nanoimprint lithography (T-NIL) is the earliest and most mature nanoimprint lithography develeoped by Professor Stephen Y. Chou's group. In a standard T-NIL process, a thin layer of imprint resist (thermal plastic polymer) is spin coated onto the sample substrate. Then the mold, which has predefined topological patterns, is brought into contact with the sample and pressed again each other under certain pressure. When heated up above the glass transition temperature of the polymer, the pattern on the mold is pressed into the melt polymer film. After being cooled down, the mold is separated from the sample and the pattern resist is left on the substrate. A pattern transfer process (Reactive Ion Etching, normally) can be used to transfer the pattern in the resist to the underneath substrate.

Step and Flash Imprint Lithography (SFIL) was developed by Prof. Grant Willson's group at the University of Texas at Austin. In SFIL, a UV curable liquid resist is applied to the sample substrate and the mold is normally made of transparent material like fused silica. After the mold and the substrate are pressed together, the resist is cured in UV light and becomes solid. After mold separation, a similar pattern transfer process can be used to transfer the pattern in resist onto the underneath material.

Nanoimprint lithography has been used to fabricate device for electrical, optical, photonic and biological applications. For electronics devices, NIL has been used to fabricate MOSFET, O-TFT, single electron memory. For optics and photonics, intensive study has been conducted in fabrication of subwavelength resonant grating filter, polarizers, waveplate, anti-reflective structures, integrated photonics circuit and plasmontic devices by NIL. sub-10 nm nanofluidic channels had been fabricated using NIL and used in DNA strenching experiment. Currently, NIL is used to shrink the size of biomolecular sorting device an order of magnitude smaller and more efficient.

A key benefit of nanoimprint lithography is its sheer simplicity. There is no need for complex optics or high-energy radiation sources. There is no need for finely tailored photoresists designed for both resolution and sensitivity at a given wavelength. The simplified requirements of the technology also lead to its low cost, another key benefit. Since large areas can be imprinted in one step, this is also a high-throughput technique.

The key concerns for nanoimprint lithography are overlay, defects, and template patterning. Due to the direct contact involved, the potential for error in overlay and potential for defects are magnified compared to cases where the image is projected from a distance. These can be mitigated with the use of effective step-and-imprint and template cleaning strategies, respectively. The current overlay 3 sigma capability is 10 nm (source). As with immersion lithography, defect control is expected to improve as the technology matures. The template patterning can currently be performed by electron beam lithography; however at the smallest resolution, the throughput is very slow. As a result, optical patterning tools will be more helpful if they have sufficient resolution. Optical patterning tools are already in use for the manufacturing of photomasks. Contact lithography or interference lithography may also be used. In the end, resolution will not be a critical factor in template generation, as a fine-resolution template (e.g., dense collection of trenches) can be formed using multiple coarse-resolution templates (e.g. a set of loosely spaced protrusions). This would lighten the burden of template generation and inspection.

A key characteristic of nanoimprint lithography is the residual layer following the imprint process. It is preferable to have thick enough residual layers to support alignment and throughput and low defects. However, this renders the nanoimprint lithography step less critical for critical dimension (CD) control than the etch step used to remove the residual layer. Hence, it is important to consider the residual layer removal an integrated part of the overall nanoimprint patterning process. In a sense, the residual layer etch is similar to the develop process in conventional lithography. It has been proposed to combine contact lithography and nanoimprint lithography techniques in one step in order to eliminate the residual layer.

A unique benefit of nanoimprint lithography is the ability to pattern 3D structures, such as damascene interconnects and T-gates, in fewer steps than required for conventional lithography. This is achieved by building the T-shape into the protrusion on the template.

Nanoimprint lithography is a simple pattern transfer process that is neither limited by diffraction nor scattering effects nor secondary electrons, and does not require any sophisticated radiation chemistry. It is also a potentially simple and inexpensive technique. However, a lingering barrier to nanometer-scale patterning is the current reliance on other lithography techniques to generate the template. It is possible that self-assembled structures will provide the ultimate solution for templates of periodic patterns at scales of 10 nm and less.

## Nanoindentation

Nanoindentation is similar to conventional hardness testing performed on a much smaller scale. The force required to press a sharp diamond indenter into a material is measured as a function of indentation depth. As depth resolution is on the scale of nanometers (hence the name of the instrument), it is possible to conduct indentation experiments even on thin films. Two quantities which can be readily extracted from nanoindentation experiments are the material's modulus, or stiffness, and its hardness, which can be correlated to yield strength. Investegators have also used nanoindentation to study creep, plastic flow, and fracture of materials.

## Nanoknot

Carbon Niv nanotubes are being used to make nanoropes; nanoknots (or nano-knot) have been tied. Science published micrograph images of some nanoknots tied from using threads made of single-wall carbon nanotube(SWNT).

It is anticipated that nanoknotting will reveal material characteristics that will not be found in unknotted nanothreads. Nano-lashings might use nanoknots to terminate as nano-items are bound together by nanomachines to construct more complex nano-items.

## Nanolithography

Nanolithography — or lithography at the nanometer scale — refers to the *fabrication of nanometer-scale structures*, meaning patterns with at least one lateral dimension between the size of an individual atom and approximately 100 nm. Nanolithography is used during the fabrication of leading-edge semiconductor integrated circuits or nanoelectromechanical systems (NEMS).

As of 2006, nanolithography is a very active area of research in academia and in industry.

Optical lithography, which has been the predominant patterning technique since the advent of the semiconductor age, is capable of producing sub-100-

nm patterns with the use of very short wavelengths (currently 193 nm). Optical lithography will require the use of liquid immersion and a host of photomask enhancement technologies (phase-shift masks (PSM), optical proximity correction (OPC)) at the 32 nm node. Most experts feel that traditional optical lithography techniques will not be cost effective below 30 nm. At that point, it may be replaced by a next-generation lithography (NGL) technique.

Other nanolithography techniques:

i) The most common nanolithographic technique is Electron-Beam Direct-Write Lithography (EBDW), the use of a beam of electrons to produce a pattern — typically in a polymeric resist such as PMMA.

ii) Extreme Ultraviolet Lithography (EUV) is a form of optical lithography using ultrashort wavelengths (13.5 nm). It is the most popularly considered NGL technique.

iii) Charged-particle lithography, such as ion- or electron-projection lithographies (PREVAIL, SCALPEL, LEEPL), are also capable of very-high-resolution patterning.

iv) Nanoimprint lithography (NIL), and its variants, such as Step-and-Flash Imprint Lithography, LISA and LADI are promising nanopattern replication technologies. This technique can be combined with contact printing.

v) Scanning Probe Lithographies (SPL) are promising tools for patterning at the deep nanometer-scale. For example, individual atoms may be manipulated using the tip of a Scanning Tunneling Microscope (STM). Dip-Pen Nanolithography (DPN) is the first commercially available SPL technology based on Atomic Force Microscopy.

vi) The furthest developed NGL remains X-ray lithography which is extensible to 15 nm resolution by use of "demagnification" in the Near Field.

## Nanomachining

Like traditional machining, where portions of the structure are removed or modified, nanomachining involves changing the structure of nano-scale materials or molecules.

## Nanomanipulation

The process of manipulating items at an atomic or molecular scale in order to produce precise structures.

## Nano Manipulator

Uses virtual reality (VR) goggles and a force feedback probe as an interface to a scanning probe microscope, providing researchers with a new way to interact with the atomic world. Researchers can travel over genes, tickle viruses, push bacteria around, and tap on molecules—the nano-Manipulator simplifies the process and allows researchers to play with their atoms. University of North Carolina at Chapel Hill (UNC-CH) *The Nanomanipulator* from the Center for Computer Integrated Systems for Microscopy and Manipulation (CISMM) at UNC Chapel Hill. Part of the Nanoscale Science Research Group (NSRG).

## Nanomaterials

Nanoscale materials; materials with structural features (particle size or grain size, for example) of at least one dimension in the range 1-100 nm. It can be subdivided into nanoparticles, nanofilms and nanocomposites. The focus of nanomaterials is a bottom up approach to structures and functional effects whereby the building blocks of materials are designed and assembled in controlled ways. Contain only a few thousand or tens of thousands of atoms, rather than the millions or billions of atoms in particles of most conventional materials.

*Nanostructures* can be obtained in metallic, ceramic, semi-conducting, and diamond materials. Applications, Nanopowders Industry. This area combines nanotechnology and many applications of nanostructured materials. One important research area is the formation of semiconductor *"quantum dots"* (i.e., several nanometer- size, faceted crystals) by injecting precursor materials conventionally used for chemical- vapor deposition of *semi-*

*conductors* into a hot liquid surfactant. This "quantum dot" is in reality a macromolecule because it is coated with a monolayer of the surfactant, preventing agglomeration.

## Nanomedicine

Nanomedicine is the medical application of nanotechnology and related research. It covers areas such as nanoparticle drug delivery and possible future applications of molecular nanotechnology (MNT) and nanovaccinology.

Current problems for nanomedicine involve understanding the issues related to toxicity and environmental impact of nanoscale materials.

Direct funding for nanomedicine projects has begun, and the US National Institute of Health received funding in 2005 to set up four nanomedicine centres. In April 2006, the journal Nature Materials estimated that 130 nanotech-based drugs and delivery systems were being developed worldwide.

The first thorough analysis of possible applications of MNT to medicine can be read in *Nanomedicine,* a book series by Robert Freitas; it analyzes a wide range of possible nanotechnology-based medical devices, and explains the relevant science behind their design.

At Rice University, a flesh welder is used to fuse two pieces of chicken meat into a single piece. The two pieces of chicken are placed together touching. A greenish liquid containing gold-coated nanoshells is dribbled along the seam. An infrared laser is traced along the seam, causing the two side to weld together. This could solve the difficulties and blood leaks caused when the surgeon tries to restitch the arteries he/she has cut during a kidney or heart transplant. The flesh welder could meld the artery into a perfect seal.

Nanoparticles of cadmium selenide (quantum dots) glow when exposed to ultraviolet light. When injected, they seep into cancer tumors. The surgeon can see the glowing tumor, and use it as a guide for more accurate tumor removal.

Jim Heath, a Caltech chemist, is developing nano-sized sensors that can detect and diagnose cancer in the early stages, when there is only a few thousand cancer cells in the body. A few drops of the patient's blood are placed on the sensor test chip. The chip contains 10,000s of nanowires that can detect proteins and other bio markers left behind by cancer cells. Cancer is curable in the early stages, so this test could save lives once perfected.

Jennifer West, a bioengineer, used nanoshells coated with gold to kill cancer tumors in mice. The nanoshells are 120 nanometers in diameter, 170 times smaller than a cancer cell. The nanoshells are injected into the mouse. The nanoshells become lodged in the cracks of the tumors. Then the mouse is shot with an infrared laser. The ray passes through the flesh harmlessly, but heats up the gold.

The gold burns the cancer cells to death, without harming the healthy cells. No mice have died, even when injected with large doses of nanoshells as per Food and Drug Administration requirement. This method is more accurate, cheaper, faster, free of side effects, and less dangerous than surgery, chemotherapy, and radiation treatment.

Largemouth bass in water containing buckyballs at 500 parts per billion suffered brain damage in 2004. Half of the lab-grown human skin and liver cells exposed to a solution, containing buckyballs at 20 parts per billion, died. Buckyballs can be made less toxic by attaching hydroxyl groups. The more hydroxyl groups added, the less toxic the buckyballs are. With the best coating, the toxicity level dropped by a factor of 10,000.

The somewhat speculative claims about the possibility of using nanorobots in medicine, advocates say, would totally change the world of medicine once it is realised. Nanomedicine would make use of these nanorobots, introduced into the body, to repair or detect damages and infections. A typical blood borne medical nanorobot would be between 0.5-3 micrometres in size, because that is the maximum size possible due to capillary passage requirement. Carbon would be the primary element used to build these nanorobots due to the inherent strength and other characteristics of some forms of carbon (diamond/fullerene composites). Cancer can be treated very effectively, according to nanomedicine advocates. Nanorobots could counter the problem of identifying and isolating cancer cells as they could be introduced into the blood stream. These nanorobots would search out cancer affected cells using certain molecular markers. Medical nanorobots would then destroy these cells, and only these cells. This could be very helpful, since current treatments like radiation therapy and chemotherapy often end up destroying more healthy cells than cancerous ones. Nanorobots could also be useful in treating vascular disease, physical trauma, and even biological aging.

### Nanomesh and Nanofibres

This term covers CNT's, and as described here, the other "nanoscale fibers" referred to as "polymeric" (made from polymers). Currently used in air and liquid filtration applications. Using a process called "electrospinning"—or e-spin—a polymer "mesh" is formed into a nanofiber membrane, hense "nanomesh", with 150—200 nm diameters. Some have been made since 1970, but were not called "nano" until recently. One potential use is *"to prevent body tissues from sticking together as they heal. It also breaks down in the body over time like biodegradable sutures."*, which makes it a surgical material for the 21st Century. Other uses include biomedical devices, filtration systems, and dust collecting systems.

### Nanometals

Metal particles of diameter in the range of a few nanometers or thin films in the same thickness range, are interesting not only because of their special mechanic properties but also because of other physical and chemical properties, sometimes totally different from those of coarse-grained metals. Metal based magnetic materials are of great interest for the field of information storage. Encapsulated (for example in a carbon matrix) metals are protected against oxidation, without loosing their magnetic properties. Encapsulated or supported nanometals are more resistant against sintering at elevated temperatures. Metallic thin films may find their application in electronic industry, for example as interconnect lines or magnetic or electric layers. Metal Based Nanomaterials, E-MRS European Materials Research Society, Fall 2004 meeting. Have a wide variety of uses including energetics (rocket propulsion and pyrotechnics), in microelectronic films and coatings, super conducting alloys and for high strength powder metallurgical metals and alloys.

Any metal that is available as a continuous ductile wire can be converted by our process into nano metal spheres. Nanomaterial Technologies, Argonide

## Nanometer (nm)

A metre (American spelling: meter; symbol: m) is a unit of length and the current base unit of length in the International System of Units (SI). The metre is part of a metric system. The metre is defined as equal to the length of the path travelled by light in absolute vacuum during a time interval of 1/299,792,458 of a second.

Historically, the metre was intended to be, and is very nearly, the ten-millionth part of the distance from the equator to the north pole. A corresponding unit of area is the square metre and a corresponding unit of volume is the cubic metre.

Multiples and subdivisions of the metre, such as *kilometre* (1000 metres) and *centimetre* (1/100 metres), are indicated by adding SI prefixes to *metre.*

The word *metre* is from the Greek *metron,* "a measure" via the French *mètre.* Its first recorded usage in English meaning this unit of length is from 1797.

In the eighteenth century, there were two favoured approaches to the definition of the standard unit of length. One suggested defining the metre as the length of a pendulum with a half-period of one second. The other suggested defining the metre as one ten-millionth of the length of the Earth's meridian along a quadrant, that is the distance from the equator to the north pole. In 1791, the French Academy of Sciences selected the meridional definition over the pendular definition because the force of gravity varies slightly over the surface of the Earth, which affects the period of a pendulum. In order to establish a universally accepted foundation for the definition of the metre, more accurate measurements of this meridian than available at that time were imperative. The Bureau des Longitudes commissioned an expedition led by Delambre and Pierre Méchain, lasting from 1792 to 1799, which measured the length of the meridian between Dunkerque and Barcelona. This portion of the meridian, which also passes through Paris, was to serve as the basis for the length of the quarter meridian, connecting the North Pole with the Equator. However, in 1793, France adopted the metre based on provisional results from the expedition as its official unit of length. Although it was later determined that the first prototype metre bar was short by a fifth of a millimetre due to miscalculation of the flattening of the Earth, this length became the standard. So, the circumference of the Earth through the poles is approximately forty million metres.

Historical *International Prototype Metre* bar, made of an alloy of platinum and iridium, that was the standard from 1889 to 1960.

In the 1870s and in light of modern precision, a series of international conferences were held to devise new metric standards. The Metre Convention (Convention du Mètre) of 1875 mandated the establishment of a permanent International Bureau of Weights and Measures (BIPM: Bureau International des Poids et Mesures) to be located in Sèvres, France. This new organization would preserve the new prototype metre and kilogram when constructed, distribute national metric prototypes, and would maintain comparisons between them and non-metric measurement standards. This organization created a new prototype bar in 1889 at the first General Conference on Weights and Measures (CGPM: Conférence Générale des Poids et Mesures), establishing the *International Prototype Metre* as the distance between two lines on a standard bar of an alloy of ninety percent platinum and ten percent iridium, measured at the melting point of ice.

In 1893, the standard metre was first measured with an interferometer by Albert A. Michelson, the inventor of the device and an advocate of using some particular wavelength of light as a standard of distance. By 1925, interferometry was in regular use at the BIPM. However, the International Prototype Metre remained the standard until 1960, when the eleventh CGPM defined the metre in the new SI system as equal to 1,650,763.73 wavelengths of the orange-red emission line in the electromagnetic spectrum of the krypton-86 atom in a vacuum. The original international prototype of the metre is still kept at the BIPM under the conditions specified in 1889.

To further reduce uncertainty, the seventeenth CGPM in 1983 replaced the definition of the metre with its current definition, thus fixing the length of the metre in terms of time and the speed of light:

## Nanomotors

A University of Florida chemistry professor has made a "nanomotor" from a single DNA molecule. The motor, so small that hundreds of thousands could fit on the head of a pin, curls up and extends like an inchworm, said Weihong Tan, the principal investigator and lead author of an article about the motor in the April edition of the journal *Nano Letters*.

## Nanonewtons

Forces 1 billion times smaller than the force required to hold an apple against Earth's gravity. Nanonewton forces are estimated with atomic force microscopes and instruments that measure the properties of ultrathin coatings like those used on computer hard drives or turbine blades. Nanotechnology: Cracking the nanonewton force barrier NIST, US, 2003.

## Nanoparticle

A nanoscale spherical or capsule-shaped structure. Most, though not all, nanoparticles are hollow, which provides a central reservoir that can be filled with anticancer drugs, detection agents, or chemicals, known as reporters, that can signal if a drug is having a therapeutic effect. The surface of a nanoparticle can also be adorned with various targeting agents, such as antibodies, drugs, imaging agents, and reporters. Most nanoparticles are constructed to be small enough to pass through blood capillaries and enter cells. The use of commercially available nanoparticle constructs for both imaging and drug delivery in humans is not novel. However, the current enthusiasm for nanotech-based solutions extends far beyond simple particles to multi-modal platforms that involve multiple active pharmaceutical ingredients (APIs). These new constructs present a higher level of complexity not only in manufacturing but also in predicting their behavior in the human including such fundamental information as bio- distribution and biocompatibility. Thus go/ no-go decision points are less well defined and decisions such as committing resources for scale up of GLP product for pre-clinical testing must be made on limited data. "The Bio-Legal Complexity of Nanoparticle Development" Dr. James L. Tatum, Special Assistant, Cancer Imaging Program, National Cancer Institute Nanoparticles, including *nano- clust*ers, [nano]- layers, [nano]- tubes, and two- and three- dimensional structures in the size range between the dimensions of molecules and 50 nm (or in a broader sense, submicron sizes as a function of materials and targeted phenomena), are seen as tailored precursors for building up functional nanostructures.

## Nanopens & Nanopencils

(AKA: Atomic Pencil) *"Analogous to using a quill pen but on a billionth the scale"*, and may transform dip-pen nanolithography. Allows for drawing electronic circuits a thousand times smaller than current ones. The "pen" is an atomic force microscope (AFM).

## NanoPGM-nanometer-scale Patterned Granular Motion

The goal of NanoPGM is to generate millions of ìnanofingers, finger-like structures each only a few nanometers long, that might someday perform precise, massively parallel manipulation of molecules and directed assembly of other nanometer-scale objects. This ability answers one of the biggest technical challenges facing builders of nanocomputers: how to arrange as many as a trillion molecular computing components in an area only a few millimeters square.

## Nanophotonics

The ongoing development of new techniques for fabricating nanoelectronic devices opens up the possibility of using these methods for making nanophotonic devices. That is, devices that involve the interaction of light with features only tens or hundreds of nanometers in size. We are currently doing exploratory research to investigate such effects and related devices that may find future use in information technology. One example is the application of interferometric lithography to the fabrication of photonic- crystal-like structures. The nanoscale physics group uses various experimental techniques to examine the physical properties of objects in the nanoscale size range, that is, a little bit larger than the size of atoms. Some interesting physical properties at this range include conductivity of small numbers of atoms and molecules, forces arising between objects on this scale, and the transition between the quantum nature of a few atoms and a large number of atoms.

## Nanopipettes

"Cantilevered/Straight Nanopipettes can be used as *nanopens* for controlled chemical delivery or removal from regions as small as 100 nanometers. They can also be used as vessels for containing molecules whose optical properties change in response to their chemical environment." Other uses include "controlled chemical etching with the precision of atomic force microscopy; chemical imaging of surfaces; delivering femtosecond laser pulses; and performing NSOM/SNOM imaging using a UV excimer laser."

## Nanoplotter

A multi-tip nanopen. "A device that can draw patterns of tiny lines just 30 molecules thick and a single molecule high. ... produces eight identical patterns at once and extends ... dip-pen nanolithography towards mass producing nanoscale devices and circuits by converting what was a serial process to a parallel one. May be use to "... miniaturize electronic circuits, pattern precise arrays of organic and biomolecules such as DNA and put thousands of different medical sensors on an area much tinier than the head of a pin."

## Nanoplumbing

Illustrations of projects that would be considered relevant to this PA. o Nanoplumbing components such as valves, microfluidic channels, and motors (e.g., to be used as pumps).

## Nanopore

Nanopore technology is an elegant concept. A membrane with very small channels (a few nanometers in diameter), called nanopores, separates two solutions. When a voltage is applied across the membrane, charged biomolecules migrate through the pores in a controlled manner.... As each nucleotide passes through the nanopore, an electronic signature is produced that can be used to characterize it.

The nanopore's size is such that it permits only a single nucleic acid strand to pass through it at any one time. Therefore the technology can be used to sequentially measure a biopolymer's properties along its length. Research in the field of *computational biology* will play a significant role in the development of nanopore technology. It will be used to help optimize experimental strategies and designs, and to create the mathematical methods needed to interpret the data generated. The potential capabilities of nanopore technology are very broad. Nanopore technology can distinguish between and count a variety of different molecules in a complex mixture. For example, nanopores could discriminate between hybridized or unhybridized unknown RNA and DNA molecules that differ by a single nucleotide only. Compared to existing techniques, nanopore technology is expected to provide direct characterization of individual nucleic acid and protein molecules directly derived from biological samples, thereby making it applicable to a wide set of analyses.

## Nanopositioners

Nanopositioners are designed to move objects in the sub-microscopic ranges of motion necessary for research work in nanotechnology. Unlike motor driven microscope stages or precision machine tools, the nanopositioner has a very limited range of motion and a correspondingly high position resolution. Total range of motion is typically limited to a few microns up to a couple of hundred microns. Absolute position can be resolved down to less than 1 nanometer (1/1000 of a micron). At this level of physical precision, care must be taken to prevent thermal expansion from adversely affecting the positioning accuracy.

## Nanoreplicators

A set of nanomachines capable of exponential replication.

## Nanorobot

A computer-controlled robotic device constructed of nanometer-scale components to molecular precision, usually microscopic in size (often abbreviated as "nanobot").

## Nanorods

or Carbon Nanorods. Formed from multi-wall carbon nanotubes. Another nanoscale material with unique and promising physical properties, such that may yield improvements in high-density data storage, and allow for cheaper flexible solar cells.

## Nanoscale

At the nanoscale, physics, chemistry, biology, materials science, and engineering converge toward the same principles and tools. The nanoscale is not just another step toward *miniaturization,* but a qualitatively new scale. The new behavior is dominated by quantum mechanics, material confinement in small structures, large interfacial volume fraction, and other unique properties, phenomena and processes. Many current theories of matter at the microscale have critical lengths of nanometer dimensions.

## Nanoscience

Nanoscience is primarily the extension of existing sciences into the realms of the extremely small (*nanomaterials, nanochemistry, nanobio, nanophysics*, etc.) while *nanoengineering* represents the extension of the engineering fields into the nano- scale realm (*nanofabrication, nanodevices*, etc.). The exponential growth of nanoscience is largely due to the development of new instruments and related techniques that are used to "routinely" probe and manipulate material at the atomic and molecular level. *Scanning probe microscopies*, analytical electron-beam techniques, epitaxial growth facilities, and *synchrotron* radiation sources are all opening huge opportunities. The prefix "nano" originates from the Greek word for dwarf, and thus refers to something small. As a prefix for a unit of time or length, it means one billionth of that unit. Thus, a nanometer (nm) is 10-9 meter. The dot over this letter "i" is approximately one million nanometers in diameter.

Nanoscience is primarily the extension of existing sciences into the realms of the extremely small (nanomaterials, nanochemistry, nanobio, nanophysics, etc.) while nanoengineering represents the extension of the engineering fields into the nano- scale realm (nanofabrication, nanodevices, etc.).

The fundamental building blocks of nature, atoms and molecules, have dimensions in the nanometer domain (i.e., the nanoscale). Many water molecules can easily occupy a sphere 1 nm in diameter. The DNA double helix is approximately 2 nm wide. The way molecules, for example, assemble into larger, supramolecular entities on the nanoscale determines important material properties (e.g., electrical, optical, and mechanical properties).

Of course, it has long been recognized that nature performs this assembly very well in the creation of the sophisticated molecular machinery that supports our life on earth. In short, by controlling structure on the scale of ~ 1 – 100 nm, one can, in principle, ultimately design new materials with specific properties. Scientists have long imagined the possibility of manipulating individual atoms and molecules.

## Nanosensor

A chemical or physical sensor constructed using nanoscale components, usually microscopic or submicroscopic in size.

## Nanoshell

Procedures that target cancer cells while leaving normal cells untouched, patients controlling the release of medicine in their bodies with an infrared light, and medical test results produced in seconds rather than days – these are three new medical technologies currently being tested by nanotechnology researchers at Rice University. ... The research focuses on nanoshells, a new type of nanoparticle invented by Naomi Halas, Rice professor of electrical and computer engineering and chemistry.

Nanoshells are layered nanoparticles whose ability to manipulate light and color can be designed into the nanoparticle by varying the thickness of the nanoparticle's layers.

## Nanosieving

In medical nanorobotics, a nanodevice that can sort molecules or other nanoscale objects by physical sieving.

## Nanospheres

Self-assembling nanospheres that fit inside each other like Russian dolls are one form of a broad range of submicroscopic spheres ... The durable silica spheres, which range in size from 2 to 50 nanometers, form in a few seconds, are small enough to be introduced into the body, and have uniform pores that could enable controlled release of drugs. The spheres can absorb organic and inorganic substances including small particles of iron, which means they can be controlled by magnets and the contents released as needed.

## Nanosprings

A nanowire wrapped into a helix. Speculation is that they "may someday make highly sensitive magnetic field detectors, perhaps finding application in hard drive read heads. Alternatively, nanosprings could serve as positioners, or even as tiny conventional springs, for nanomachines of the future."

## Nanostructures

Nanometer sized objects. Nanostructures may be considered as small, familiar, or large, depending on the view point of the disciplines concerned. To chemists, nanostructures are molecular assemblies of atoms numbering from $10^3$ to $10^9$ and of molecular weights of $10^4$ to $10^{10}$ Daltons. Thus, they are chemically large supramolecules. To molecular biologists, nanostructures have the size of familiar objects from proteins to viruses and cellular organelles. But to material scientists and electrical engineers, nanostructures are the current limit of microfabrication and thus are rather small. Nanostructures are complex systems which evidently lie at the interface between solid- state physics, supramolecular chemistry, and molecular biology.

### Nanosystem

A eutactic set of nanoscale components working together to serve a set of purposes; complex nanosystems can be of macroscopic size.

### Nanotechnology

Nanotechnology is the design, characterization, production and application of structures, devices and systems by controlling shape and size at the nanoscale. Eight to ten atoms span one nanometer (nm). The human hair is approximately 70,000 to 80,000 nm thick.

Nanotechnology should really be called "nanotechnologies": There is no single field of nanotechnology. The term broadly refers to such fields as biology, physics or chemistry, any scientific field, or a combination thereof, that deals with the deliberate and controlled manufacturing of nanostructures.The United States' National Nanotechnology Initiative website defines it as follows: "Nanotechnology is the understanding and control of matter at dimensions of roughly 1 to 100 nanometers, where unique phenomena enable novel applications."

Nanoscience is the study of phenomena and manipulation of material at the nanoscale, in essence an extension of existing sciences into the nanoscale. Nanoscience is the world of atoms, molecules, macromolecules, quantum dots, and macromolecular assemblies, and is dominated by surface effects such as Van der Waals force attraction, hydrogen bonding, electronic charge, ionic bonding, covalent bonding, hydrophobicity, hydrophilicity, and quantum mechanical tunneling, to the virtual exclusion of macro-scale effects such as turbulence and inertia. For example, the vastly increased ratio of surface area to volume opens new possibilities in surface-based science, such as catalysis.

The ongoing quest for miniaturization has resulted in tools such as the atomic force microscope (AFM) and the scanning tunneling microscope (STM). Combined with refined processes such as electron beam lithography, these instruments allow us to deliberately manipulate and manufacture nanostructures. Engineered nanomaterials, either by way of a top-down approach (a bulk material is reduced in size to nanoscale pattern) or a bottom-up approach (larger structures are built or grown atom by atom or molecule by molecule), go beyond just a further step in miniaturization. They have broken a size barrier below which quantization of energy for the electrons in solids becomes relevant. The so-called "quantum size effect" describes the physics of electron properties in solids with great reductions in particle size. This effect does not come into play by going from macro to micro dimensions. However, it becomes dominant when the nanometer size range is reached. Materials reduced to the nanoscale can suddenly show very different properties compared to what they show on a macroscale. For instance, opaque substances become transparent (copper); inert materials become catalysts (platinum); stable materials turn combustible (aluminum); solids turn into liquids at room temperature (gold); insulators become conductors (silicon). A second important aspect of the nanoscale is that the smaller a nanoparticle gets, the larger its relative surface area becomes. Its electronic structure changes dramatically, too. Both effects lead to greatly improved catalytic activity but can also lead to aggressive chemical reactivity.

The fascination with nanotechnology stems from these unique quantum and surface phenomena that matter exhibits at the nanoscale, making possible novel applications and interesting materials.

*History of Use*

The first mention of some of the distinguishing concepts in nanotechnology (but predating use of that name) was in "There's Plenty of Room at the Bottom," a talk given by physicist Richard Feynman at an American Physical Society meeting at Caltech on December 29, 1959. Feynman described a process by which the ability to manipulate individual atoms and molecules might be developed, using one set of precise tools to build and operate another proportionally smaller set, so on down to the needed scale. In the course of this, he noted, scaling issues would arise from the changing magnitude of various physical phenomena: gravity would become less important, surface tension and Van der Waals attraction would become more important, etc. This basic idea appears feasible, and exponential assembly enhances it with parallelism to produce a useful quantity of end products.

The term "nanotechnology" was defined by Tokyo Science University Professor Norio Taniguchi in a 1974 paper (N. Taniguchi, "On the Basic Concept of 'Nano-Technology'," Proc. Intl. Conf. Prod. Eng. Tokyo, Part II, Japan Society of Precision Engineering, 1974.) as follows: "'Nano-technology' mainly consists of the processing of, separation, consolidation, and deformation of materials by one atom or one molecule." In the 1980s the basic idea of this definition was explored in much more depth by Dr. Eric Drexler, who promoted the technological significance of nano-scale phenomena and devices through speeches and the books Engines of Creation: The Coming Era of Nanotechnology and *Nanosystems: Molecular Machinery, Manufacturing, and Computation,* (ISBN 0-471-57518-6), and so the term acquired its current sense.

More broadly, nanotechnology includes the many techniques used to create structures at a size scale below 100 nm, including those used for fabrication of nanowires, those used in semiconductor fabrication such as deep ultraviolet lithography, electron beam lithography, focused ion beam machining, nanoimprint lithography, atomic layer deposition, and molecular vapor deposition, and further including molecular self-assembly techniques such as those employing di-block copolymers. However, all of these techniques preceded the nanotech era, and are extensions in the development of scientific advancements rather than techniques which were devised with the sole purpose of creating nanotechnology or which were results of nanotechnology research.

Nanotechnology and nanoscience got started in the early 1980s with two major developments; the birth of cluster science and the invention of the scanning tunneling microscope (STM). This development led to the discovery of fullerenes in 1986 and carbon nanotubes a few years later. In another development, the synthesis and properties of semiconductor nanocrystals was studied. This led to a fast increasing number of metal oxide nanoparticles of quantum dots.

Technologies currently branded with the term 'nano' are little related to and fall far short of the most ambitious and transformative technological goals of the sort in molecular manufacturing proposals, but the term still connotes such ideas. Thus there may be a danger that a "nano bubble" will form from the use of the term by scientists and entrepreneurs to garner funding, regardless of (and perhaps despite a lack of) interest in the transformative possibilities of more ambitious and far-sighted work. The diversion of support based on the promises of proposals like molecular manufacturing to more mundane projects also risks creating a perhaps unjustifiedly cynical impression of the most ambitious goals: an investor intrigued by molecular manufacturing who invests in 'nano' only to find typical materials science advances result might conclude that the whole idea is hype, unable to appreciate the bait-and-switch made possible by the vagueness of the term. On the other hand, some have argued that the publicity and competence in related areas generated by supporting such 'soft nano' projects is valuable, even if indirect, progress towards nanotechnology's most ambitious goals.

In describing nanostructures we need to differentiate between the number of dimensions of the nanoscale. Nanotextured surfaces are one dimension on the nanoscale, i.e., only the thickness of the surface of an object is between 0.1 and 100 nm. Nanotubes are two dimensions on the nanoscale, i.e., the diameter of the tube is between 0.1 and 100nm; its length could be much greater. Finally, spherical nanoparticles are three dimensions on the nanoscale, i.e., the particle is between 0.1 and 100 nm in each spatial dimension. The terms nanoparticles and ultrafine particles (UFP) often are used synonymously although UFP can reach into the micron range.

Nanoscience and nanotechnology only became possible in the 1980s with the development of the first tools to measure and make nanostructures.

The atomic force microscope (AFM) and the Scanning Tunneling Microscope (STM) are two early versions of scanning probes that launched nanotechnology. There are other types of scanning probe microscopy, all based on the idea of the STM, that make it possible to see structures at the nanoscale. The tip of scanning probes can also be used to manipulate nanostructures (a process called positional assembly). However, this is a very slow process. This led to the development of various techniques of nanolithography such as dip pen nanolithography, electron beam lithography or nanoimprint lithography. Lithography is a top-down fabrication technique where a bulk material is reduced in size to nanoscale pattern.

In contrast, bottom-up techniques build or grow larger structures atom by atom or molecule by molecule. These techniques include chemical synthesis, self-assembly and positional assembly.

With nanotechnology, a large set of materials with distinct properties (optical, electrical, or magnetic) can be fabricated. Nanotechnologically improved products rely on a change in the physical properties when the feature sizes are shrunk. Nanoparticles for example take advantage of their dramatically increased surface area to volume ratio. Their optical properties, e.g. fluorescence, become a function of the particle diameter. When brought into a bulk material, nanoparticles can strongly influence the mechanical properties, such as the stiffness or elasticity. Example, traditional polymers can be reinforced by nanoparticles resulting in novel materials e.g. as lightweight replacements for metals. Therefore, an increasing societal benefit of such nanoparticles can be expected.

Such nanotechnologically enhanced materials will enable a weight reduction accompanied by an increase in stability and an improved functionality.

The biological and medical research communities have exploited the unique properties of nanomaterials for various applications (e.g., contrast agents for cell imaging and therapeutics for treating cancer). Terms such as biomedical nanotechnology, bionanotechnology, and nanomedicine are used to describe this hybrid field.

Functionalities can be added to nanomaterials by interfacing them with biological molecules or structures. The size of nanomaterials is similar to that of most biological molecules and structures; therefore, nanomaterials can be useful for both in vivo and in vitro biomedical research and applications. Thus far, the integration of nanomaterials with biology has led to the development of diagnostic devices, contrast agents, analytical tools, therapy, and drug-delivery vehicles.

Nanotechnology-on-a-chip is one more dimension of lab-on-a-chip technology. Biological tests measuring the presence or activity of selected substances become quicker, more sensitive and more flexible when certain nanoscale particles are put to work as tags or labels. Magnetic nanoparticles, bound to a suitable antibody, are used to label specific molecules, structures or microorganisms. Gold nanoparticles tagged with short segments of DNA can be used for detection of genetic sequence in a sample. Multicolor optical coding for biological assays has been achieved by embedding different-sized quantum dots into polymeric microbeads. Nanopore technology for analysis of nucleic acids converts strings of nucleotides directly into electronic signatures.

Chemical catalysis benefits especially from nanoparticles, due to the extremely large surface to volume ratio. The application potential of nanoparticles in catalysis ranges from fuel cell to catalytic converters and photocatalytic devices. Catalysis is also important for the production of chemicals.

The overall drug consumption and side-effects can be lowered significantly by depositing the active agent in the morbid region only and in no higher dose than needed. This highly selective approach reduces costs and human suffering. An example can be found in dendrimers and nanoporous materials. They could hold small drug molecules transporting them to the desired location. Another vision is based on small electromechanical systems: NEMS are being investigated for the active release of drugs. Some potentially important applications include cancer treatment with iron nanoparticles or gold shells.

A targeted or personalized medicine reduces the drug consumption and treatment expenses resulting in an overall societal benefit by reducing the costs to the public health system. -u-uu/-u.

Nanotechnology can help to reproduce or to repair damaged tissue. This so called "tissue engineering" makes use of artificially stimulated cell proliferation by using suitable nanomaterial-based scaffolds and growth factors. Tissue engineering might replace today's conventional treatments, e.g. transplantation of organs or artificial implants. On the other hand, tissue engineering is closely related to the ethical debate on human stem cells and its ethical implications.

Chemical catalysis and filtration techniques are two prominent examples where nanotechnology already plays a role. The synthesis provides novel materials with tailored features and chemical properties e.g. nanoparticles with a distinct chemical surrounding (ligands) or specific optical properties. In this sense, chemistry is indeed a basic nanoscience. In a short-term perspective, chemistry will provide novel "nanomaterials" and in the long run, superior processes such as "self-assembly" will enable energy and time preserving strategies.

In a sense, all chemical synthesis can be understood in terms of nanotechnology, because of its ability to manufacture certain molecules. Thus, chemistry forms a base for nanotechnology providing tailor-made molecules, polymers etc. and furthermore clusters and nanoparticles.

A strong influence of nanochemistry on waste-water treatment, air purification and energy storage devices is to be expected. Mechanical or chemical methods can be used for effective filtration techniques. One class of filtration techniques is based on the use of membranes with suitable hole sizes, whereby the liquid is pressed through the membrane. Nanoporous membranes are suitable for a mechanical filtration with extremely small pores smaller than 10 nm Nanofiltration is mainly used for the removal of ions or the separation of different fluids. On a larger scale, the membrane filtration technique is named ultrafiltration, which works down to between 10 and 100 nm. One important field of application for ultrafiltration is medical purposes as can be found in renal dialysis.

Magnetic nanoparticles offer an effective and reliable method to remove heavy metal contaminants from waste water by making use of magnetic separation techniques. Using nanoscale particles increases the efficiency to absorb the contaminants and is comparatively inexpensive compared to traditional precipitation and filtration methods.

The most advanced nanotechnology projects related to energy are: storage, conversion, manufacturing improvements by reducing materials and process rates, energy saving e.g. by better thermal insulation, and enhance renewable energies sources.

A reduction of energy consumption can be reached by better insulation systems, by the use of more efficient lighting or combustion systems, and by use of lighter and stronger materials in the transportation sector. Currently used light bulbs only convert approximately 5% of the electrical energy into light. Nanotechnological approaches like LEDs (Light-emitting diodes) or QCAs (Quantum Caged Atoms) could lead to a strong reduction of energy consumption for illumination.

Today's best solar cells have layers of two different semiconductors stacked together to absorb light at different energies but they still only manage to use 30 percent of the Sun's energy. Commercially available solar cells have much lower efficiencies (less than 20%). Nanotechnology can help to increase the efficiency of light conversion by specifically designed nanostructures. The degree of efficiency of combustion engines is not higher than 15-20% at the moment. Nanotechnology can improve combustion by designing specific catalysts with maximized surface area.

An example for an environmentally friendly form of energy is the use of fuel cells powered by hydrogen, which is ideally produced by renewable energies. Probably the most prominent nanostructured material in fuel cells is the catalyst consisting of carbon supported noble metal particles with diameters of 1- 5 nm. Suitable materials for hydrogen storage contain a large number of small nanosized pores. Therefore many nanostructured materials like nanotubes, zeolites or alanates are under investigation.

Nanotechnology can contribute to the further reduction of combustion engine pollutants by nanoporous filters, which can clean the exhaust mechanically, by catalytic converters based on nanoscale noble metal particles or by catalytic coatings on cylinder walls and catalytic nanoparticles as additive for fuels.

Because of the relatively low energy density of batteries the operating time is limited and a replacement or recharging is needed. The huge number of spent batteries and accumulators represent a disposal problem. The use of batteries with higher energy content or the use of rechargeable batteries or supercapacitors with higher rate of recharging using nanomaterials could be helpful for the battery disposal problem.

Current high-technology production processes are based on traditional top down strategies, where nanotechnology has already been introduced silently. The critical length scale of integrated circuits is already at the nanoscale (50 nm and below) regarding the gate length of transistors in CPUs or DRAM devices.

An example of such novel devices is based on spintronics.The dependence of the resistance of a material (due to the spin of the electrons) on an external field is called magnetoresistance. This effect can be significantly amplified (GMR – Giant Magneto-Resistance) for nanosized objects, for example when two ferromagnetic layers are separated by a nonmagnetic layer, which is several nanometers thick (e.g. Co-Cu-Co). The GMR effect has led to a strong increase in the data storage density of hard disks and made the gigabyte range possible. The so called tunneling magnetoresistance (TMR) is very similar to GMR and based on the spin dependant tunneling of electrons through adjacent ferromagnetic layers. Both the GMR- and the TMR-effect can be used to create a non-volatile main memory for computers, such as the so called magnetic random access memory or MRAM.

In the modern communication technology traditional analog electrical devices are increasingly replaced by optical or optoelectronic devices due to their enormous bandwidth and capacity, respectively. Two promising examples are photonic crystals and quantum dots.

Photonic crystals are materials with a periodic variation in the refractive index with a lattice constant that is half the wavelength of the light used. They offer a selectable band gap for the propagation of a certain wavelength, thus they resemble a semiconductor, but for light or photons instead of electrons.

Quantum dots are nanoscaled objects, which can be used, among many other things, for the construction of lasers. The advantage of a quantum dot laser over the traditional semiconductor laser is that their emitted wavelength depends on the diameter of the dot. Quantum dot lasers are cheaper and offer a higher beam quality than conventional laser diodes.

The production of displays with low energy consumption could be accomplished using carbon nanotubes. CNTs can be electrically conductive and due to their small diameter of several nanometers, they can be used as field emitters with extremely high efficiency for field emission displays (FED). The principle of operation resembles that of the cathode ray tube, but on a much smaller length scale.

Entirely new approaches for computing exploit the laws of quantum mechanics for novel quantum computers, which enable the use of fast quantum algorithms.

Nanotechnology is already impacting the field of consumer goods, providing products with novel functions ranging from easy-to-clean to scratch-resistant. Modern textiles are wrinkle-resistant and stain-repellent; in the mid-term clothes will become "smart", through embedded "wearable electronics".

Already in use are different nanoparticle improved products. Especially in the field of cosmetics such novel products have a promising potential.

Nanotechnology can be applied in the production, processing, safety and packaging of food. A nanocomposite coating process could improve food packaging by placing anti-microbial agents directly on the surface of the coated film. Nanocomposites could increase or decrease gas permeability of different fillers as is needed for different products. They can also improve the mechanical and heat-resistance properties and lower the oxygen transmission rate. Research is being performed to apply nanotechnology to the detection of chemical and biological substances for sensing biochemical changes in foods.

The most prominent application of nanotechnology in the household is self-cleaning or "easy-to-clean" surfaces on ceramics or glasses. Nanoceramic particles have improved the smoothness and heat resistance of common household equipment such as the flat iron.

The first sunglasses using protective and antireflective ultrathin polymer coatings are on the market. For optics, nanotechnology also offers scratch resistant coatings based on nanocomposites.

The use of engineered nanofibers already makes clothes water- and stain-repellent or wrinkle-free. Textiles with a nanotechnological finish can be washed less frequently and at lower temperatures. Nanotechnology has been used to integrate tiny carbon particles membrane and guarantee full-surface protection from electrostatic charges for the wearer.

Tennis rackets with carbon nanotubes have an increased torsion and flex resistance. The rackets are more rigid than current carbon rackets and pack more power. Long-lasting tennis-balls are made by coating the inner core with clay polymer nanocomposites. These tennis-balls have twice the lifetime of conventional balls.

One field of application is in sunscreens. The traditional chemical UV protection approach suffers from its poor long-term stability. A sunscreen based on mineral nanoparticles such as titanium dioxide offer several advantages. Titanium dioxide nanoparticles have a comparable UV protection property as the bulk material, but lose the cosmetically undesirable whitening as the particle size is decreased.

The mere presence of nanomaterials (materials that contain nanoparticles) is not in itself a threat. It is only certain aspects that can make them risky, in particular their mobility and their increased reactivity. Only if certain properties of certain nanoparticles were harmful to living beings or the environment would we be faced with a genuine hazard.

In addressing the health and environmental impact of nanomaterials we need to differentiate two types of nanostructures: (i) "free" nanoparticles, where at some stage in production or use individual nanoparticles of a substance are present and (ii) Nanocomposites, nanostructured surfaces and nanocomponents (electronic, optical, sensors etc.), where nanoscale particles are incorporated into a substance, material or device ("fixed" nano-particles). These free nanoparticles could be nanoscale species of elements, or simple compounds, but also complex compounds where for instance a nanoparticle of a particular element is coated with another substance ("coated" nanoparticle or "core-shell" nanoparticle).

There seems to be consensus that, although one should be aware of materials containing fixed nanoparticles, the immediate concern is with free nanoparticles.

Because nanoparticles are very different from their everyday counterparts, their adverse effects cannot be derived from the known toxicity of the macro-sized material. This poses significant issues for addressing the health and environmental impact of free nanoparticles.

To complicate things further, in talking about nanoparticles it is important that a powder or liquid containing nanoparticles is almost never monodisperse, but will contain a range of particle sizes. This complicates the experimental analysis as larger nanoparticles might have different properties than smaller ones. Also, nanoparticles show a tendency to aggregate and such aggregates often behave differently from individual nanoparticles.

There are four entry routes for nanoparticles into the body: they can be inhaled, swallowed, absorbed through skin or be deliberately injected during medical procedures (or released from implants). Once within the body they are highly mobile and in some instances can even cross the blood-brain barrier.

How these nanoparticles behave inside the organism is one of the big issues that needs to be resolved. Basically, the behavior of nanoparticles is a function of their size, shape and surface reactivity with the surrounding tissue. They could cause "overload" on phagocytes, cells that ingest and destroy foreign matter, thereby triggering stress reactions that lead to inflammation and weaken the body's defense against other pathogens. Apart from what happens if non- or slowly degradable nanoparticles accumulate in organs, another concern is their potential interaction with biological processes inside the body: because of their large surface, nanoparticles on exposure to tissue and fluids will immediately absorb onto their surface some of the macromolecules they encounter. Can this, for instance, affect the regulatory mechanisms of enzymes and other proteins?

Certain nanoparticles can also have negative effects on the body. For example, in a lake, trout have been known to die of overexposure to the slightest amount of Carbon 60 (a molecule consiting of 60 carbon atoms, known technically as buckminsterfullerenes or colloquially as "bucky balls"). The bucky ball itself is a popular molecule in the production of carbon based nanostructures such as carbon tubes.

In fixed form, where they are part of a manufactured substance or product, they will ultimately have to be recycled or disposed of as waste. We don't know yet if certain nanoparticles will constitute a completely new class of non-biodegradable pollutant. In case they do, we also don't know yet how such pollutants could be removed from air or water because most traditional filters are not suitable for such tasks (their pores are too big to catch nanoparticles).

Health and environmental issues combine in the workplace of companies engaged in producing or using nanomaterials and in the laboratories engaged in nanoscience and nanotechnology research. It is safe to say that current workplace exposure standards for dusts cannot be applied directly to nanoparticle dusts.

To properly assess the health hazards of engineered nanoparticles the whole life cycle of these particles needs to be evaluated, including their fabrication, storage and distribution, application and potential abuse, and disposal. The impact on humans or the environment may vary at different stages of the life cycle.

An often cited worst-case scenario is "grey goo", a hypothetical substance into which the surface of the earth might be transformed by self-replicating nanobots running amok. This concept has been analyzed by Freitas in "Some Limits to Global Ecophagy by Biovorous Nanoreplicators, with Public Policy Recommendations". With the advent of nan-biotech, a different scenario called green goo has been forwarded. Here, the malignant substance is not nanobots but rather self-replicating organisms engineered through nanotechnology.

Societal risks from the use of nanotechnology have also been raised. On the instrumental level, these include the possibility of military applications of nanotechnology (such as implants and other means for soldier enhancement, such as those being developed at the Institute for Soldier Nanotechnologies at MIT) as well as enhanced surveillance capabilities through nano-sensors.

On the structural level, critics of nanotechnology point to a new world of ownership and corporate control opened up by nanotechnology. The claim is that, just as biotechnology's ability to manipulate genes went hand in hand with the patenting of life, so too nanotechnology's ability to manipulate molecules has led to the patenting of matter. The last few years has seen a gold rush to claim patents at the nanoscale. Over 800 nano-related patents were granted in 2003, and the numbers are increasing year to year. Corporations are already taking out broad ranging monopoly patents on nanoscale discoveries and inventions. For example, two corporations, NEC and IBM, hold the basic patents on carbon nanotubes, one of the current cornerstones of nanotechnology. Carbon nanotubes have a wide range of uses, and look set to become crucial to several industries from electronics and computers, to strengthened materials to drug delivery and diagnostics. Carbon nanotubes are poised to become a major traded commodity with the potential to replace major conventional raw materials. However, as their use expands, anyone seeking to manufacture or sell carbon nanotubes, no matter what the application, must first buy a license from NEC or IBM.

Regulatory bodies such as the Environmental Protection Agency and the Food and Drug Administration in the U.S. or the Health & Consumer Protection Directorate of the European Commission have started dealing with the potential risks posed by nanoparticles. So far, neither engineered nanoparticles nor the products and materials that contain them are subject to any special regulation regarding production, handling or labeling. The Material Safety Data Sheet that must be issued for certain materials often do not differentiate between bulk and nanoscale size of the material in question.

Studies of the health impact of airborne particles are the closest thing we have to a tool for assessing potential health risks from free nanoparticles. These studies have generally shown that the smaller the particles get, the more toxic they become. This is due in part to the fact that, given the same mass per volume, the dose in terms of particle numbers increases as particle size decreases. Looking at all available data, it must be concluded that current risk assessment methodologies are not suited to the hazards associated with nanoparticles; in particular, existing toxicological and eco-toxicological methods are not up to the task; exposure evaluation (dose) needs to be expressed as quantity of nanoparticles and/or surface area rather than simply mass; equipment for routine detecting and measuring nanoparticles in air, water or soil is inadequate; and very little is known about the physiological responses to nanoparticles.

Regulatory bodies in the U.S. as well as in the EU have concluded that nanoparticles form the potential for an entirely new risk and that it is necessary to carry out an extensive analysis of the risk. The outcome of these studies can then form the basis for government and international regulations.

When the term "nanotechnology" was independently coined and popularized by Eric Drexler, who at the time was unaware of Taniguchi's usage, it referred to a future manufacturing technology based on molecular machine systems. The premise was that molecular-scale biological analogies of traditional machine components demonstrated that molecular machines were possible, and that a manufacturing technology based on the mechanical functionality of these components (such as gears, bearings, motors, and structural members) would enable programmable, positional assembly to atomic specification. The physics and engineering performance of exemplar designs were analyzed in the textbook Nanosystems. Because the term "nanotechnology" was subsequently applied to other uses, new terms evolved to refer to this distinct usage: "molecular nanotechnology," "molecular manufacturing," and most recently, "productive nanosystems."

One alternative view is that designs such as those proposed by Drexler and Merkle do not accurately account for the electrostatic interactions and will not operate according to the results of the analysis in Nanosystems. The contention is that man-made nanodevices will probably bear a much stronger resemblance to other (less mechanical) nanodevices found in nature: cells, viruses, and prions.

Another view, put forth by Carlo Montemagno, is that future nanosystems will be hybrids of silicon technology and biological molecular machines, and his group's research is directed toward this end.

The seminal experiment proving that positional molecular assembly is possible was performed by Ho and Lee at Cornell University in 1999. They used a scanning tunneling microscope to move an individual carbon monoxide molecule (CO) to an individual iron atom (Fe) sitting on a flat silver crystal, and chemically bind the CO to the Fe by applying a voltage.

Though biology clearly demonstrates that molecular machine systems are possible, non-biological molecular machines are today only in their infancy. Leaders in research on non-biological molecular machines are Dr. Alex Zettl and his colleagues at Lawrence Berkeley Laboratories and UC Berkeley. They have constructed at least three distinct molecular devices whose motion is controlled from the desktop with changing voltage: a rotating molecular motor, a molecular actuator, and a nanoelectromechanical relaxation oscillator. Manufacturing in the context of productive nanosystems is not related to, and should be clearly distinguished from, the conventional technologies used to manufacture nanomaterials such as carbon nanotubes and nanoparticles.

One of the problems facing nanotechnology concerns how to assemble atoms and molecules into smart materials and working devices. Supramolecular chemistry, a very important tool here, is the chemistry beyond the molecule, and molecules are being designed to self-assemble into larger structures. In this case, biology is a place to find inspiration: cells and their pieces are made from self-assembling biopolymers such as proteins and protein complexes. One of the things being explored is synthesis of organic molecules by adding them to the ends of complementary DNA strands such as ——A and ——B, with molecules A and B attached to the end; when these are put together, the complementary DNA strands hydrogen bonds into a double helix, ====AB, and the DNA molecule can be removed to isolate the product AB.

Advanced nanotechnology, sometimes called molecular manufacturing, is a term given to the concept of engineered nanosystems (nanoscale machines) operating on the molecular scale. By the countless examples found in biology it is currently known that billions of years of evolutionary feedback can produce sophisticated, stochastically optimized biological machines, and it is hoped that developments in nanotechnology will make possible their construction by some shorter means, perhaps using biomimetic principles. However, K Eric Drexler and other researchers have proposed that advanced nanotechnology, although perhaps initially implemented by biomimetic means, ultimately could be based on mechanical engineering principles.

In August 2005, a task force consisting of 50+ international experts from various fields was organized by the Center for Responsible Nanotechnology to study the societal implications of molecular nanotechnology. Determining a set of pathways for the development of molecular nanotechnology is now an objective of a broadly based technology roadmap project led by Battelle (the manager of several U.S. National Laboratories) and the Foresight Institute. That roadmap should be completed by early 2007.

A definitive feature of nanotechnology is that it constitutes an interdisciplinary ensemble of several fields of the natural sciences that are, in and of themselves, actually highly specialized. Thus, physics play an important role—alone in the construction of the microscope used to investigate such phenomena but above all in the laws of quantum mechanics. Chemistry also play an important role in the identification of the materials, steps and synthesis of molecules for nanotechnical devices. It is not surprising, then, since Physics and Chemistry share the same principles and overlap so much, that Physical Chemistry (or Chemical physics) is probably the most fundamental discipline in the nanotechology research. This is an important concept to mention, because a lot of research depends on the background that future students will have. Of course, the best educational pathway to this type of research (perhaps the only one to master this source) is a multidisciplinary one, one with basic science and engineering knowledge.

## Nanotubes

Nanometer-sized tubes composed of various substances including carbon ( CARBON NANOTUBES), boron nitride, or nickel vanadate. A one dimensional *fullerene* (a convex cage of atoms with only hexagonal and/ or pentagonal faces) with a cylindrical shape. Carbon nanotubes discovered in 1991 by Sumio Iijima resemble rolled up graphite, although they can not really be made that way. Depending on the direction that the tubes appear to have been rolled (quantified by the 'chiral vector'), they are known to act as conductors or semiconductors. Nanotubes are a proving to be useful as molecular components for nanotechnology.

### Nanowire

A nanometer-scale wire made of metal atoms, silicon, or other materials that conduct electricity. Nanowires are built atom by atom on a solid surface, often as part of a microfluidic device. They can be coated with molecules such as antibodies that will bind to proteins and other substances of interest to researchers and clinicians. By the very nature of their nanoscale size, nanowires are incredibly sensitive to such binding events and respond by altering the electrical current flowing through them, and thus can form the basis of ultra sensitive molecular detectors.

### Nastic Structures

The Nastic Materials Program is exploring the development of a new class of active materials that can mimic the ability of plants to generate large strains while still performing a structural function. The impetus for this work is the desire to develop a highly controllable and reversible material system that can generate 10 Megapascals and 20 percent in blocked stress and free strain, respectively.

The ultimate goal is the development of plant- inspired actuation systems that possess the power density of conventional hydraulic systems for application in military adaptive or morphing structures.

### National Nanotechnology Initiative

US federal government agencies participating include the National Science Foundation, the Department of Defense, the National Institute of Health, NASA, and NIST.

### Navigation Satellite

A type of satellite that gives ships and aircraft their coordinate positions on the Earth. Navigation satellites were developed in the 1950's, and they rely on the *doppler effect* to calculate the position of vessels emitting a radio signal. Navigation satellites are also widely used by the military.

### Navstar

A series of American *navigation satellites.* The first satellite in the Navstar system was launched in 1978. Currently, there are 24 orbiting Navstar satellites. The three-dimensional satellite navigational system of Navstar enables a traveller to find out his or her position anywhere on or above the planet.

### Neologomania

The transhuman enthusiasm for neologisms, especially the creation of words with many prefixes.

### Neomorph

A structure, part, or organ developed independently, that is, not derived from a similar structure, part, or organ, in a preexisting form.

### Neophile

The trait of being excited and pleased by novelty. Common among most hackers, SF fans, and members of several other connected leading-edge subcultures, including the pro-technology 'Whole Earth' wing of the ecology movement, space activists, many members of Mensa, and the Discordian/ neo-pagan underground. All these groups overlap heavily and (where evidence is available) seem to share characteristic hacker tropisms for science fiction, music, and oriental food. The opposite tendency is 'neophobia'.

### Neutrino

A particle with very low mass (around that of an electron), and no electrical charge, the neutrino is an elusive subatomic particle. The neutrino is so shy that the duration between the theorization of its existence and its actual discovery was 25 years. Wolfgang Pauli, a famous quantum physicist, theorized the neutrino in 1931. It was discovered by Frederick Reines and Clyde Cowan in 1956 at a neutrino observatory located adjacent to a nuclear power plant at Savannah River, South Carolina. Neutrinos travel at almost the speed of light, and many quadrillions of them penetrate your body every second. But because neutrinos have such low mass and interact only slightly with atoms, they can penetrate several light years of densely packed matter before interacting with an atom. For this reason they are very difficult to detect.

Neutrinos are generated during an event known in physics as beta decay. It seemed hopeless to detect neutrinos until the advent of nuclear technology. Atomic bombs and nuclear reactors proved to be rich sources of neutrino activity relative to a typical spot on Earth. The first neutrino detectors were tanks filled with water and cadmium chloride. The first neutrino detected was in fact not a conventional neutrino but an anti-neutrino. When an anti-neutrino collided with a proton in the neutrino detector, the interaction produced a neutron and a positron, or an anti-electron.

The resulting anti-electron would quickly annihilate with one of the electrons orbiting the nucleus, resulting in a spray of two photons. Then a stray neutron released from the breakdown of the atom would eventually (~15ms) be picked up by another, intact atom, releasing more photons (light). This distinct 2-stage pattern of photon release could be magnified by photoamplifiers, thereby triggering a register and providing positive evidence for neutrino impact. With modern methods, as many as one neutrino per day is detected in our observatories.

The neutrino is an excellent example of a fundamental particle that becomes more understandable as the quality of our scientific instruments improves. The continued gathering of evidence regarding the neutrino and its properties is sure to contribute in valuable ways to the progress of contemporary theoretical physics, which in turn will generate useful technological and theoretical discoveries for human civilization.

## Nonlinear

Advances in genomic technologies are a mix of incremental improvements to existing technologies (linear) and occasionally, a truly new paradigm or breakthrough. Related terms: disruptive technologies, emerging technologies, complex.

## Nuclear Magnetic Resonance

An analytical method that can detect subatomic and structural information of molecules by measuring the adsorption of radio-frequency electromagnetic radiation by nuclei under the influence of a magnetic field.

## Nuclear Properties

By definition, any two atoms with an identical number of *protons* in their nuclei belong to the same chemical element. Atoms with equal numbers of protons but a different number of *neutrons* are different isotopes of the same element. For example, all hydrogen atoms admit exactly one proton, but isotopes exist with no neutrons (hydrogen-1, by far the most common form, sometimes called protium), one neutron (deuterium), two neutrons (tritium) and more than two neutrons. The known elements form a set of atomic numbers from hydrogen with a single proton up to the 118-proton element ununoctium. All known isotopes of elements with atomic numbers greater than 82 are radioactive.

About 339 nuclides occur naturally on Earth, of which 269 (about 79%) are stable. Of the chemical elements, 80 have one or more stable isotopes. Elements 43, 61, and all elements numbered 83 or higher have no stable isotopes. As a rule, there is, for each atomic number (each element) only a handful of stable isotopes, the average being 3.4 stable isotopes per element which has any stable isotopes. Sixteen elements have only a single stable isotope, while the largest number of stable isotopes observed for any element is ten (for the element tin).

Stability of isotopes is affected by the ratio of protons to neutrons, and also by presence of certain "magic numbers" of neutrons or protons which represent closed and filled quantum shells. These quantum shells correspond to a set of energy levels within the shell model of the nucleus. Of the 269 known stable nuclides, only four have both an odd number of protons *and* odd number of neutrons: H, Li, B and N. Also, only four naturally-occurring, radioactive odd-odd nuclides have a half-life over a billion years: K, V, La and Ta. Most odd-odd nuclei are highly unstable with respect to beta decay, because the decay products are even-even, and are therefore more strongly bound, due to nuclear pairing effects.

## Nucleus

All the bound protons and neutrons in an atom make up a tiny atomic nucleus, and are collectively called nucleons. The radius of a nucleus is approximately equal to fm, where *A* is the total number of nucleons. This is much smaller than the radius of the atom, which is on the order of 10 fm. The nucleons are bound together by a short-ranged attractive potential called the residual strong force. At distances smaller than 2.5 fm this force is much more powerful than the electrostatic force that causes positively charged protons to repel each other. Atoms of the same element have the same number of protons, called the atomic number. Within a single element, the number of neutrons may vary, determining the isotope of that element.

The total number of protons and neutrons determine the nuclide. The number of neutrons relative to the protons determines the stability of the nucleus, with certain isotopes undergoing radioactive decay. The neutron and the proton are different types of fermions. The Pauli exclusion principle is a quantum mechanical effect that prohibits *identical* fermions (such as multiple protons) from occupying the same quantum physical state at the same time. Thus every proton in the nucleus must occupy a different state, with its own energy level, and the same rule applies to all of the neutrons. A nucleus that has a different number of protons than neutrons can potentially drop to a lower energy state through a radioactive decay that causes the number of protons and neutrons to more closely match.

As a result, atoms with roughly matching numbers of protons and neutrons are more stable against decay. However, with increasing atomic number, the mutual repulsion of the protons requires an increasing proportion of neutrons to maintain the stability of the nucleus, which modifies this trend.

Thus, there are no stable nuclei with equal proton and neutron numbers above atomic number Z = 20 (calcium); and as Z increases toward the heaviest nuclei, the ratio of neutrons per proton required for stability increases to about 1.5. The number of protons and neutrons in the atomic nucleus can be modified, although this can require very high energies because of the strong force.

Nuclear fusion occurs when multiple atomic particles join to form a heavier nucleus, such as through the energetic collision of two nuclei. At the core of the Sun, protons require energies of 3–10 KeV to overcome their mutual repulsion—the coulomb barrier—and fuse together into a single nucleus. Nuclear fission is the opposite process, causing a nucleus to split into two smaller nuclei—usually through radioactive decay. The nucleus can also be modified through bombardment by high energy subatomic particles or photons. In such processes that change the number of protons in a nucleus, the atom becomes an atom of a different chemical element.

If the mass of the nucleus following a fusion reaction is less than the sum of the masses of the separate particles, then the difference between these two values is emitted as energy, as described by Albert Einstein's mass–energy equivalence formula, $E = mc$, where $m$ is the mass loss and $c$ is the speed of light. This deficit is the binding energy of the nucleus.

# O

## Oligodendrocyte

Oligodendrocytes (from Greek literally meaning *cells with a few branches*), or oligodendroglia (Greek, *few tree glue*), are a variety of neuroglia. Their main function is the insulation of the axons exclusively in the central nervous system of the higher vertebrates, a function performed by Schwann cells in the peripheral nervous system.

A single oligodendrocyte can extend to up to 50 axons, wrapping around approximately 1 mm of each and forming the myelin sheath; Schwann cells, on the other hand, can only wrap around 1 axon.

Oligodendroglia arise during development from an oligodendrocyte precursor cell, which can be identified by its expression of a number of antigens, including the ganglioside GD3, the NG2 chondroitin sulfate proteoglycan, and the platelet-derived growth factor-alpha receptor subunit PDGF-alphaR. In the rat forebrain the majority of oligodendroglial progenitors arise during late embryogenesis and early postnatal development from cells of the subventricular zones (SVZ) of the lateral ventricles. SVZ cells migrate away from these germinal zones to populate both developing white and gray matter, where they differentiate and mature into myelin-forming oligodendroglia. However, it is not clear whether all oligodendroglial progenitors undergo this sequence of events. It has been suggested that some undergo apoptosis and that some fail to differentiate into oligodendroglia but persist into maturity as adult oligodendroglial progenitors.

The nervous system of mammals depends crucially on the myelin sheath for insulation as it results in decreased ion leakage and lower capacitance of the cell membrane. There is also an overall increase in impulse speed as saltatory propagation of action potentials occurs at the nodes of Ranvier in between Schwann cells (of the PNS) and oligodendrocytes (of the CNS); furthermore miniaturization occurs, whereby impulse speed of myelinated axons increases linearly with the axon diameter, whereas the impulse speed of unmyelinated cells increases only with the square root of the diameter.

As part of the nervous system they are closely related to nerve cells and like all other glial cells the oligodendrocytes have a supporting role towards neurons. They are intimately involved in signal propagation, providing the same functionality as the insulation on a household electrical wire (with the rather large difference that while household electrical wires are in a non-conducting medium - air - the axons run in a solution of water and ions that conduct electrical current well).

Satellite oligodendrocytes are functionally distinct from most oligodendrocytes. They are not attached to neurons and therefore do not serve an insulating role. They remain close to neurons and regulate the extracellular fluid.

Diseases that result in injury to the oligodendroglial cells include demyelinating diseases such as multiple sclerosis and leukodystrophies. Cerebral palsy (periventricular leukomalacia) is caused by damage to developing oligodendrocytes in the brain areas around the cerebral ventricles. Spinal cord injury also causes damage to oligodendrocytes. In cerebral palsy, spinal cord injury, stroke and possibly multiple sclerosis, oligodendrocytes are thought to be damaged by excessive release of the neurotransmitter glutamate.

Oligodendrocyte dysfunction may also be implicated in the pathophysiology of schizophrenia and bipolar disorder. Oligodendroglia are also susceptible to infection by the JC virus, which causes progressive multifocal leukoencephalopathy (PML), a condition which specifically affects white matter, typically in immunocompromised patients. Tumors of oligodendroglia are called oligodendrogliomas.

## Oligomer

Short polymer, usually consisting (in a cell) of amino acids (oligopeptides), sugars (oligosaccharides), or nucleotides (oligonucleotides). (From Greek *oligos*, few, little.)

## Omega Point

Also called the Quantum Omega Point Theory. A possible future state when intelligence controls the Universe totally, and the amount of information processed and stored goes asymptotically towards infinity.

## Orbital Tower

Also known as a "space tether", "beanstalk" or "heavenly funicular". A cable in synchronous orbit, with one end anchored to the surface of the Earth, often with a small asteroid at the outer end to provide some extra tension and stability. Picture also a "space elevator". In theory, constructed of a diamondoid material, approximately 22,000 miles long, with one end in a stable orbit, and the other somewhere around the equator. Used frequently in science-fiction yarns, and may become a reality with the advent of mature MNT.

Such an elevator would move freight and passengers into orbit at a cost per pound orders of magnitude less than current launches, with passenger safety comparable to train, plane, or subway trips. Becomes possible when we can mass-produce nanotubes, and make their length to fit.

## Orbital

In the approximation that each electron in a molecule has a distinct, independent wave function, the spatial distribution of an electron wave function corresponds to a molecular orbital. These, in turn, can be approximated as sums of contributions from the orbitals characteristic of the isolated atoms. An electron added to a molecule—or, similarly, one excited to a higher-energy state within a molecule—would occupy a state with a different wave function from the rest; an unoccupied state of this kind corresponds to an unoccupied molecular orbital. Orbital-symmetry effects on reaction rates arise when a reaction requires overlap between two lobes of the orbitals on each of two reagents: if the algebraic signs of the wave functions in the facing lobes do not match, bond formation between those orbitals is prohibited.

## Organic Moledule

A molecule containing carbon; the complex molecules in living systems are all organic molecules in this sense.

## Organic Synthesis

Combinatorial libraries & synthesis Narrower terms high throughput organic synthesis

## Organotypic

Despite their wide use, the physiological relevance of organotypic slices remains controversial. Such cultures are prepared at 5 days postnatal. Although some local circuitry remains intact, they develop subsequently in isolation from the animal and hence without plasticity due to experience.

Development of synaptic connectivity and morphology might be expected to proceed differently under these conditions than in a behaving animal.

## Oscillator

Circuit that produces an alternating voltage (current) when supplied by a steady (DC) energy source.

## Outmessaging

In medical nanorobotics, conveyance of information from a transmitter located inside the human body, especially from working nanodevices, to the patient or to a recipient external to the human body.

## Overcoming Cytoxicity

Studies have shown that quantum dot toxicity is due to cadmium ions being released into cells. This is exacerbated by oxygen, UV exposure and the large exposed surfaces of the spherical quantum dots.

To avoid these problems, surfaces are added to the quantum dots, traditionally consisting of Zinc-Sulfide compounds to create a more benign surface. Polymers are used as well, which have been shown to be stable.

## Overlap

Orbitals lack sharply defined surfaces, declining in amplitude exponentially in their surface regions. When two orbitals are brought together, regions of substantial amplitude overlap. The resulting system can be described as two new orbitals, one formed by joining the two original orbitals without introducing a node in the wave function, and the other formed with a node between them.

The nodeless joining reduces the energy of the electrons relative to the separate orbitals, resulting in a bonding interaction; joining with a node raises the energy, producing an antibonding interaction. If both new orbitals are occupied, antibonding forces dominate, resulting in overlap repulsion. Molecular mechanics models give an approximate description of overlap (and other) forces for a certain range of atoms and geometries.

## Oxidation

Oxidation (Redox) reactions include all chemical processes in which atoms have their oxidation number (oxidation state) changed. This can be a simple redox process, such as the oxidation of carbon to yield carbon dioxide, it could be the reduction of carbon by hydrogen to yield methane, or it could be the oxidation of sugar in the human body, through a series of very complex electron transfer processes. The term *redox* comes from the two concepts of reduction and oxidation. It can be explained in simple terms: Oxidation describes the loss of an electron by a molecule, atom or ion Reduction describes the gain of an electron by a molecule, atom or ion However, these descriptions are not truly correct. Oxidation and reduction properly refer to *a change in oxidation number*—the actual transfer of electrons may never occur. Thus, oxidation is better defined as an *increase in oxidation number*, and reduction as a *decrease in oxidation number*. In practice, the transfer of electrons will always cause a change in oxidation number, but there are many reactions which are classed as "redox", though no electrons are transferred.

# P

## Peroxisome

Small membrane-bounded organelle that uses molecular oxygen to oxidize organic molecules. Contains some enzymes that produce and others that degrade hydrogen peroxide ($H_2O_2$).

## Phagocytosis

Process by which particulate material is endocytosed ("eaten") by a cell. Prominent in carnivorous cells, such as *Amoeba proteus,* and in vertebrate macrophages and neutrophils. (From Greek *phagein,* to eat.)

## Pharmacodelivery

Site-directed pharmacodelivery is a desirable but elusive goal. Endothelium and epithelium create formidable barriers to endogenous molecules as well as targeted therapies *in vivo.*

## Pharmacophore Generation

A procedure to extract the most important common structural features relevant for a given biological activity from a series of molecules with a similar mechanism of action.

## Pharmacyte

In medical nanorobotics, a theorized (nanorobotic) device capable of delivering precise doses of biologically active chemicals to individually-addressed human body tissue cells (e.g. cell-by-cell drug delivery).

## Pharmainformatics

The multidisciplinary informatics needs of the pharmaceutical industry (HTS High Throughput Screening) data, combinatorial chemistry, ADME informatics, cheminformatics, toxicology, etc. information access and communication between various departments like the development and discovery teams.

## Phase Space

A classical system of $N$ particles can be described by its $3N$ position and $3N$ momentum coordinates. The phase space associated with the system is the $6N$ dimensional space defined by these coordinates.

## Phonon

A particle of sound. The energy $E$ of a phonon is given by the Einstein relation, $E = hf$. Here $f$ is the frequency of the sound and $h$ is Planck's constant. The momentum $p$ of a photon is given by the de Broglie relation, $p = h/ë$. Here ë is the wavelength of the sound. A quantum of acoustic energy, analogous to the quantum of electromagnetic radiation, the photon. Thermal excitations in a crystal or in an elastic continuum can be described as a population of phonons (analogous to blackbody electromagnetic radiation). In highly inhomogeneous solids, a description in terms of phonons breaks down and localized vibrational modes become important.

## Phosphorescence

Luminescence that occurs at times greater than on the order of a second after an electron excitation event.

## Photolithographic Mask

A template used in photolithography that allows selective exposure of a photosensitive surface.

## Photon

A particle of light. The energy $E$ of a photon is given by the Einstein relation, $E = hf$. Here $f$ is the frequency of the light and $h$ is Planck's constant. The momentum $p$ of a photon is given by the de Broglie relation, $p = h/ë$. Here ë is the wavelength of the light.

## Photonic Band Gap (PBG)

An energy range (and corresponding wavelength range) for which a material neither absorbs light nor allows light propagation.

## Photonic Crystal

Composite material made as periodical arrangement of pieces of material (e.g. spheres) embedded in a background matrix of another material that has very different (much lower) dielectric constant and/or magnetic permittivity (electromagnetic properties of material). Such structures do not allow for propagation of light of specific frequencies, due to a destructive interference of photons of such frequencies (this is a wave phenomenon also known as photonic band gap).

Photonic crystals offer a possibility of manipulating the emission, propagation and splitting of electromagnetic waves which is expected to be useful for technologies of all-photon telecommunications. The production of such materials relies on nanotechnological methods (e.g. lithography, self assembly approaches...).

## Phototransistor

Transistor that, when powered, produces amplified voltage (current) in response to illumination.

## Pi bond

A covalent bond formed by overlap between two *p* orbitals on different atoms. Pi bonds are superimposed on sigma bonds, forming double or triple bonds.

## Pico Technology

(trillionth of a meter) — The next step smaller, after Nanotechnology. The art of manipulating materials on a quantum scale.

## Piezoelectric Material

A ferroelectric material in which an electrical potential difference is created due to mechanical deformation, or conversely, in which the application of a voltage causes dimensional changes in the material.

## Pinhole

The term pinhole embraces a wide variety of oxide defects and is used in a broad sense today. Listed in this category are cracks caused by thermal contraction after oxidation or by handling, and regions of oxide with low dielectric strength caused by dust particles, inadequate masking, contamination, or poor resist adhesion.

## Pinhole

The term pinhole embraces a wide variety of oxide defects and is used in a broad sense today. Listed in this category are cracks caused by thermal contraction after oxidation or by handling, and regions of oxide with low dielectric strength caused by dust particles, inadequate masking, contamination, or poor resist adhesion.

### Pink Goo

(humorous) Humans (in analogy with grey goo). "Pink Goo to refer to Old Testament apes who see their purpose as being fruitful and multiplying, filling up of the cosmos with lots more such apes, unmodified."

### Pin-out

Diagram showing for electronic components the relations between connecting pins and internal components.

### Pitting

A form of very localized corrosion wherein small pits or holes form, usually in a vertical direction.

### pK value

A measure of the strength of an acid on a logarithmic scale. The pK value is given by log10 (1/Ka), where Ka is the acid dissociation constant pK values often are used to compare the strengths of different acids.

A low molecular weight polymer additive that enhances flexibility and workability and reduces stiffness and brittleness.

### Point Defect

A crystalline defect associated with one or several atomic sites.

### Poisson's Ratio

A bar of an isotropic, elastic material ordinarily shrinks laterally when it is stretched longitudinally. The lateral contracting strain divided by the applied tensile strain is Poisson's ratio, which varies from material to material.

### Polar Orbit

Usually has an angle of inclination of 90 degrees to the equator. On every pass around the Earth, it passes over both the north and south poles. Therefore, as the Earth rotates to the east underneath the satellite which is travelling north and south, it can cover the entire Earth's surface. A polar orbiting satellite covers the entire globe every 14 days.

### Polar

Satellite launched on February 24, 1996 by NASA in the Global Geospace Science project. Polar is an atmospheric studies satellite in polar orbit. One purpose of Polar is to gather information that will help scientists protect future satellites from radiation and other atmospheric dangers.

### Polycyclic

A cyclic structure contains rings of bonds; a structure having many such rings is termed polycyclic. In the polycyclic structures of interest in this volume, a large fraction of the atoms are members of multiple small rings, resulting in considerable rigidity.

### Polyimide

A family of polymers involving carbon and nitrogen bonds, known for good thermal stability.

### Polymeric Materials

Materials composed of large molecules, generally based on carbon, that have been formed from the chemical bonding of smaller units (monomers).

### Polymerization

A chemical reaction resulting in the bonding together of small molecular units (monomers) to form a much larger molecule (polymer). There are many different routes, or reaction mechanisms, for polymerization.

### Polymers-biomedical

More and more therapeutic problems are relevant to the use of polymer- based therapeutic aids for a limited period of time, namely the healing time related to the outstanding capacity of living systems to self-repair, e.g. bone fracture fixation with screws and plates, of wound closure by sutures and also of drug delivery from implants or similar systems based on polymeric matrices, or on aqueous dispersions or solutions of polymers.

### Positional Navigation

In medical nanorobotics, a form of nanorobotic navigation in which nanodevices know their exact location inside the human body to ~micron accuracy continuously at all times.

## Positional Synthesis

Control of chemical reactions by precisely positioning the reactive molecules; the basic principle of assemblers.

## Positive Sum

A term used to describe a situation in which one or more entities can gain without other entities suffering an equal loss—for example, a growing economy.

## POSS Nanotechnology

Short for Polyhedral Oligomeric Silsesquioxanes Nanotechnology. POSS nanomaterials are attractive for missile and satellite launch rocket applications because they offer effective protection from collisions with space debris and the extreme thermal environments of deep space and atmospheric re-entry.

Another application of POSS nanotechnology under development is a new high-temperature lubricant. This new nanolubricant is effective at temperatures up to 500ƒF, which is 100ƒF greater than conventional lubricants. From Technologies developed by the Propulsion Directorate's Polymer Working Group at Edwards AFB

## Post Monetary Economy

After the advent of mature Nanotechnology, it is likely that our economic reality will change, possibly to the extent of eliminating currency as we know it today.

## Posthuman

Persons of unprecedented physical, intellectual, and psychological capacity, self-programming, self-constituting, potentially immortal, unlimited individuals.

## Potential Energy Surface

The potential energy of a ground-state molecular system containing $N$ atoms is a function of its geometry, defined by $3N$ spatial coordinates (a configuration space). If the energy is imagined as corresponding to a height in a $3N + 1$ dimensional space, the resulting landscape of hills, hollows, and valleys is the potential energy surface.

## Potential Energy

Energy stored in an object due to its position or internal

structure. The energy associated with a configuration of particles, as distinct from their motions. In macroscopic experience, potential energy can be increased (for example) by stretching a spring or by lifting a mass against a gravitational force; in molecular systems, potential energy can be increased (for example) by stretching a bond or by separating molecules against a van der Waals attraction.

### Power [W]

Product of voltage and current in a component; also, refers to the field of electric energy supply.

### Precision

The degree of reproducibility among several independent measurements of the same true value under specified conditions.

### Presentation Semaphore

In medical nanorobotics, a mechanical device used to display specific antigens, chemical ligands, or other molecular objects to the external environment, with the purpose of selectively modifying the chemical or other surface characteristics of a nanorobot exterior.

### Privileged Structures

The retrospective analysis of the chemical structure of various drugs used in medicine led medicinal chemists to identify some molecular motifs that are more frequently associated with higher biological activity than other structures. Such molecular motifs were called "privileged" structures by Evans et al., to mean substructures that confer activity to two or more different receptors. The implication was that the privileged structure provides the scaffold, and that substitutions on the scaffold provide specificity for a particular receptor. Two monographs describe the privileged structure concept. Historically, the most popular privileged structures were the arylethylamines (including indolylethylamines), the diphenylmethane derivatives, the tricyclic psychotropics, and the sulfonamides. The dihydropyridines, the benzodiazepines, the N-arylpiperazines, and the biphenyls are more recent contributions.

## Probability Density Function

Consider an uncertain physical property and a corresponding space describing the range of values that the property can have (e.g., the configuration of a thermally excited $N$ particle system and the corresponding $3N$ dimensional configuration space). The probability density function associated with a property is defined over the corresponding space; its value at a particular point is the probability per unit volume that the property has a value in an infinitesimal region around that point.

## Protein Design, Protein Engineering

The design and construction of new proteins; an enabling technology for nanotechnology.

## Protein Folding

"The process by which proteins acquire their functional, preordained, three-dimensional structure after they emerge, as linear polymers of amino acids, from the ribosome."

## Protein

Living cells contain many molecules that consist of amino acid polymers folded to form more-or-less definite three-dimensional structures; these are termed proteins. Short polymers lacking definite three-dimensional structures are termed peptides.

Many proteins incorporate structures other than amino acids, either as covalently attached side chains or as bound ligands. Molecular objects made of protein form much of the molecular machinery of living cells.

## Proteomics

The study of the full expression of proteins in an organism The term proteome refers to all the proteins expressed by a genome, and thus proteomics involves the identification of proteins in the body and the determination of their role in physiological and pathophysiological functions. ... Ultimately it is believed that through proteomics new disease markers and drug targets can be identified that will help design products to prevent.

## Proximal Probes

A family of devices capable of fine positional control and sensing, including scanning tunneling and atomic force microscopes; an enabling technology for nanotechnology.

# Q

## Quantum Computer

A quantum computer is a device for computation that makes direct use of distinctively quantum mechanical phenomena, such as superposition and entanglement, to perform operations on data. In a classical (or conventional) computer, information is stored as bits; in a quantum computer, it is stored as qubits (quantum binary digits). The basic principle of quantum computation is that the quantum properties can be used to represent and structure data, and that quantum mechanisms can be devised and built to perform operations with these data.

Although quantum computing is still in its infancy, experiments have been carried out in which quantum computational operations were executed on a very small number of qubits. Both practical and theoretical research continues with interest, and many national government and military funding agencies support quantum computing research to develop quantum computers for both civilian and national security purposes, such as cryptanalysis. If large-scale quantum computers can be built, they will be able to solve certain problems much faster than any of our current classical computers (for example Shor's algorithm).

Quantum computers are different from other computers such as DNA computers and traditional computers based on transistors. Some computing architectures such as optical computers may use classical superposition of electromagnetic waves. Without some specifically quantum mechanical resources such as entanglement, it is conjectured that an exponential advantage over classical computers is not possible. A classical computer has a memory made up of bits, where each bit holds either a one or a zero. A quantum computer maintains a sequence of qubits. A single qubit can hold a one, a zero, or, crucially, a quantum superposition of these; moreover, a pair of qubits can be in a quantum superposition of 4 states, and three qubits in a superposition of 8. In general a quantum computer with n qubits can be in up to $2^n$ different states simultaneously (this compares to a normal computer that can only be in *one* of these $2^n$ states at any one time). A quantum computer operates by manipulating those qubits with a fixed sequence of quantum logic gates. The sequence of gates to be applied is called a *quantum algorithm.*

While a classical three-bit state and a quantum three-qubit state are both eight-dimensional vectors, they are manipulated quite differently for classical or quantum computation, respectively. For computing in either case, the system must be initialized, for example into the all-zeros string, i.e., (1,0,0,0,0,0,0,0) or , In classical randomized computation, the system evolves according to the application of stochastic matrices, which preserve that the probabilities add up to one (i.e., preserve the L1 norm). In quantum computation, on the other hand, allowed operations are unitary matrices, which are effectively rotations (they preserve that the sum of the squares add up to one, the Euclidean or L2 norm). (Exactly what unitaries can be applied depend on the physics of the quantum device.) Consequently, since rotations can be undone by rotating backward, quantum computations are reversible. (Technically, quantum operations can be probabilistic combinations of unitaries, so quantum computation really does generalize classical computation. See quantum circuit for a more precise formulation.)

Finally, upon termination of the algorithm, the result needs to be read off. In the case of a classical computer, we *sample* from the probability distribution on the three-bit register to obtain one definite three-bit string, say 000. Quantumly, we *measure* the three-qubit state, which is equivalent to collapsing the quantum state down to a classical distribution (with the coefficients in the classical state being the squared magnitudes of the coefficients for the quantum state, as described above) followed by sampling from that distribution. Note that this destroys the original quantum state. Many algorithms will only give the correct answer with a certain probability, however by repeatedly initializing, running and measuring the quantum computer, the probability of getting the correct answer can be increased. For example, running the Shor factorization algorithm four times will give the correct answer with a very high probability.

For more details on the sequences of operations used for various algorithms, see universal quantum computer, Shor's algorithm, Grover's algorithm, Deutsch-Jozsa algorithm, quantum Fourier transform, quantum gate, quantum adiabatic algorithm and quantum error correction.

Integer factorization is believed to be computationally infeasible with an ordinary computer for large integers that are the product of only a few prime numbers (e.g., products of two 300-digit primes). By comparison, a quantum computer could efficiently solve this problem using Shor's algorithm to find its factors. This ability would allow a quantum computer to "break" many of the cryptographic systems in use today, in the sense that there would be a polynomial time (in the number of bits of the integer) algorithm for solving the problem. In particular, most of the popular public key ciphers are based on the difficulty of factoring integers (or the related discrete logarithm problem which can also be solved by Shor's algorithm), including forms of RSA. These are used to protect secure Web pages, encrypted email, and many other types of data. Breaking these would have significant ramifications for electronic privacy and security. The only way to increase the security of an algorithm like RSA would be to increase the key size and hope that an adversary does not have the resources to build and use a powerful enough quantum computer. A way out of this dilemma would be to use some kind of quantum cryptography. There are also some digital signature schemes that are believed to be secure against quantum computers. See for instance Lamport signatures.

Besides factorization and discrete logarithms, quantum algorithms offering a more than polynomial speedup over the best known classical algorithm have been found for several problems, including the simulation of quantum physical processes from chemistry and solid state physics, the approximation of Jones polynomials, and solving Pell's equation. There is no mathematical proof that an equally fast classical algorithm cannot be discovered, although this is considered unlikely. For some problems, quantum computers offer a polynomial speedup. The most well-known example of this is *quantum database search*, which can be solved by Grover's algorithm using quadratically fewer queries to the database than are required by classical algorithms. In this case the advantage is provable. Several other examples of provable quantum speedups for query problems have subsequently been discovered, such as for finding collisions in two-to-one functions and evaluating NAND trees.

An example of this is a password cracker that attempts to guess the password for an encrypted file (assuming that the password has a maximum possible length). For problems with all four properties, the time for a quantum computer to solve this will be proportional to the square root of $n$ (it would take an average of $(n + 1)/2$ guesses to find the answer using a classical computer.) That can be a very large speedup, reducing some problems from years to seconds. It can be used to attack symmetric ciphers such as Triple DES and AES by attempting to guess the secret key. Regardless of whether any of these problems can be shown to have an advantage on a quantum computer, they nonetheless will always have the advantage of being an excellent tool for studying quantum mechanical interactions, which of itself is an enormous value to the scientific community.

Grover's algorithm can also be used to obtain a quadratic speed-up for a class of problems known as NP-complete.

There are a number of practical difficulties in building a quantum computer, and thus far quantum computers have only solved trivial problems. David DiVincenzo, of IBM, listed the following requirements for a practical quantum computer: (i) scalable physically to increase the number of qubits; (ii) qubits can be initialized to arbitrary values; (iii) quantum gates faster than decoherence time; (iv) universal gate set; (v) qubits can be read easily. To summarize the problems from the perspective of an engineer, one needs to solve the challenge of building a system which is isolated from everything *except* the measurement and manipulation mechanism. Furthermore, one needs to be able to turn off the coupling of the qubits to the measurement so as to not decohere the qubits while performing operations on them.

One major problem is keeping the components of the computer in a coherent state, as the slightest interaction with the external world would cause the system to decohere. This effect causes the unitary character (and more specifically, the invertibility) of quantum computational steps to be violated. Decoherence times for candidate systems, in particular the transverse relaxation time $T_2$ (terminology used in NMR and MRI technology, also called the *dephasing time*), typically range between nanoseconds and seconds at low temperature. These issues are more difficult for optical approaches as the timescales are orders of magnitude lower and an often cited approach to overcoming them is optical pulse shaping. Error rates are typically proportional to the ratio of operating time to decoherence time, hence any operation must be completed much more quickly than the decoherence time.

If the error rate is small enough, it is thought to be possible to use quantum error correction, which corrects errors due to decoherence, thereby allowing the total calculation time to be longer than the decoherence time. An often cited (but rather arbitrary) figure for required error rate in each gate is $10^{-4}$. This implies that each gate must be able to perform its task 10,000 times faster than the decoherence time of the system.

Meeting this scalability condition is possible for a wide range of systems. However, the use of error correction brings with it the cost of a greatly increased number of required qubits. The number required to factor integers using Shor's algorithm is still polynomial, and thought to be between $L$ and $L^2$, where $L$ is the number of bits in the number to be factored; error correction algorithms would inflate this figure by an additional factor of $L$. For a 1000-bit number, this implies a need about $10^4$ qubits without error correction. With error correction, the figure would rise to about $10^7$ qubits. Note that computation time is about $L^2$ or about $10^7$ steps and on 1 MHz, about 10 seconds.

A very different approach to the stability-decoherence problem is to create a topological quantum computer with anyons, quasi-particles used as threads and relying on braid theory to form stable logic gates.

## Quantum Cryptography

Quantum cryptography, or quantum key distribution (QKD), uses quantum mechanics to guarantee secure communication. It enables two parties to produce a shared random bit string known only to them, which can be used as a key to encrypt and decrypt messages.

An important and unique property of quantum cryptography is the ability of the two communicating users to detect the presence of any third party trying to gain knowledge of the key. This results from a fundamental part of quantum mechanics: the process of measuring a quantum system in general disturbs the system. A third party trying to eavesdrop on the key must in some way measure it, thus introducing detectable anomalies.

By using quantum superpositions or quantum entanglement and transmitting information in quantum states, a communication system can be implemented which detects eavesdropping. If the level of eavesdropping is below a certain threshold a key can be produced which is guaranteed as secure (i.e. the eavesdropper has no information about), otherwise no secure key is possible and communication is aborted.

The security of quantum cryptography relies on the foundations of quantum mechanics, in contrast to traditional public key cryptography which relies on the computational difficulty of certain mathematical functions, and cannot provide any indication of eavesdropping or guarantee of key security.

Quantum cryptography is only used to produce and distribute a key, not to transmit any message data. This key can then be used with any chosen encryption algorithm to encrypt (and decrypt) a message, which can then be transmitted over a standard communication channel. The algorithm most commonly associated with QKD is the one-time pad, as it is provably secure when used with a secret, random key.

Quantum communication involves encoding information in quantum states, or qubits, as opposed to classical communications use of bits. Usually, photons are used for these quantum states. Quantum cryptography exploits certain properties of these quantum states to ensure its security. There are several different approaches to quantum key distribution, but they can be divided into two main categories depending of which property they exploit.

This protocol, known as BB84 after its inventors and year of publication, was originally described using photon polarization states to transmit the information. However any two pairs of conjugate states can be used for the protocol, and many optical fibre based implementations described as BB84 use phase encoded states. The sender (traditionally referred to as Alice) and the receiver (Bob) are connected by a quantum communication channel which allows quantum states to be transmitted. In the case of photons this channel is generally either an optical fibre or simply free space. In addition they communicate via a public classical channel, for example using radio waves or the internet. Neither of these channels need to be secure; the protocol is designed with the assumption that an eavesdropper (referred to as Eve) can interfere in any way with both.

The security of the protocol comes from encoding the information in non-orthogonal states. Quantum indeterminacy means that these states cannot in general be measured without disturbing the original state (see No cloning theorem). BB84 uses two pairs of states, with each pair conjugate to the other pair, and the two states within a pair orthogonal to each other. Pairs of orthogonal states are referred to as a basis. The usual polarization state pairs used are either the rectilinear basis of vertical (0°) and horizontal (90°), the diagonal basis of 45° and 135° or the circular basis of left- and right-handedness. Any two of these bases are conjugate to each other, and so any two can be used in the protocol. Below the rectilinear and diagonal bases are used.

The first step in BB84 is quantum transmission. Alice creates a random bit (0 or 1) and then randomly selects one of her two bases (rectilinear or diagonal in this case) to transmit it in. She then prepares a photon polarization state depending both on the bit value and basis, as shown in the table to the left. So for example a 0 is encoded in the rectilinear basis (+) as a vertical polarization state, and a 1 is encoded in the diagonal basis (x) as a 135° state. Alice then transmits a single photon in the state specified to Bob, using the quantum channel. This process is then repeated from the random bit stage, with Alice recording the state, basis and time of each photon sent.

Quantum mechanics (particularly quantum indeterminacy) says there is no possible measurement that will distinguish between the 4 different polarization states, as they are not all orthogonal. The only measurement possible is between any two orthogonal states (a basis), so for example measuring in the rectilinear basis will give a result of horizontal or vertical. If the photon was created as horizontal or vertical (as a rectilinear eigenstate) then this will measure the correct state, but if it was created as 45° or 135° (diagonal eigenstates) then the rectilinear measurement will instead return either horizontal or vertical at random. Furthermore, after this measurement the photon will be polarized in the state it was measured in (horizontal or vertical), with all information about its initial polarization lost.

As Bob does not know the basis the photons were encoded in, all he can do is select a basis at random to measure in, either rectilinear or diagonal. He does this for each photon he receives, recording the time, measurement basis used and measurement result. After Bob has measured all the photons, he communicates with Alice over the public classical channel. Alice broadcasts the basis each photon was sent in, and Bob the basis each was measured in. They both discard photon measurements (bits) where Bob used a different basis, which will be half on average, leaving half the bits as a shared key.

To check for the presence of eavesdropping Alice and Bob now compare a certain subset of their remaining bit strings. If a third party has gained any information about the photons polarization it will have introduced errors in Bobs measurements. If more than $p$ bits differ they abort the key and try again, possibly with a different quantum channel, as the security of the key cannot be guaranteed. $p$ is chosen so that if the number of bits known to Eve is less than this, privacy amplification can be used to reduce Eve's knowledge of the key to an arbitrarily small amount, by reducing the length of the key.

The quantum cryptography protocols described above will provide Alice and Bob with nearly identical shared keys, and also with an estimate of the discrepancy between the keys. These differences can be caused by eavesdropping, but will also be caused by imperfections in the transmission line and detectors. As it is impossible to distinguish between these two types of errors, it is assumed all errors are due to eavesdropping in order to guarantee security. Provided the error rate between the keys is lower than a certain threshold (20% as of April 2007), two steps can be performed to first remove the erroneous bits and then reduce Eve's knowledge of the key to an arbitrary small value. These two steps are known as information reconciliation and privacy amplification respectively, and were first described in 1992.

Information reconciliation is a form of error correction carried out between Alice and Bob's keys, in order to ensure both keys are identical. It is conducted over the public channel and as such it is vital to minimise the information sent about each key, as this can be read by Eve. A common protocol used for information reconciliation is the cascade protocol, proposed in 1994. This operates in several rounds, with both keys divided into blocks in each round and the parity of those blocks compared. If a difference in parity is found then a binary search is performed to find and correct the error. If an error is found in a block from a previous round that had correct parity then another error must be contained in that block; this error is found and corrected as before. This process is repeated recursively, which is the source of the cascade name. After all blocks have been compared, Alice and Bob both reorder their keys in the same random way, and a new round begins. At the end of multiple rounds Alice and Bob will have identical keys with high probability, however Eve will have gained additional information about the key from the parity information exchanged.

Privacy Amplification is a method for reducing (and effectively eliminating) Eve's partial information about Alice and Bob's key. This partial information could have been gained both by eavesdropping on the quantum channel during key transmission (thus introducing detectable errors), and on the public channel during information reconciliation (where it is assumed Eve gains all possible parity information). Privacy amplification uses Alice and Bob's key to produce a new, shorter key, in such a way that Eve has only negligible information about the new key. This can be done using a universal hash function, chosen at random from a publicly known set of such functions, which takes as its input a binary string of length equal to the key and outputs a binary string of a chosen shorter length. The amount by which this new key is shortened is calculated, based on how much information Eve could have gained about the old key (which is known due to the errors this would introduce), in order to reduce the probability of Eve having any knowledge of the new key to a very low value.

As of March 2007 the longest distance over which quantum key distribution has been demonstrated using optic fibre is 148.7 km, achieved by Los Alamos/NIST using the BB84 protocol. Significantly, this distance is long enough for almost all the spans found in today's fibre networks. The distance record for free space QKD is 144km between two of the Canary Islands, achieved by a European collaboration using entangled photons (the Ekert scheme) in 2006, and using BB84 enhanced with decoy states in 2007. The experiments suggest transmission to satellites is possible, due to the lower atmospheric density at higher altitudes. For example although the minimum distance from the International Space Station to the ESA Space Debris Telescope is about 400 km, the atmospheric thickness is about an order of magnitude less than in the European experiment, thus yielding less attenuation compared to this experiment.

The DARPA Quantum Network, a 10-node quantum cryptography network, has been running since 2004 in Massachusetts, USA. It is being developed by BBN Technologies, Harvard University, Boston University and QinetiQ.

There are currently three companies offering commercial quantum cryptography systems; id Quantique (Geneva), MagiQ Technologies (New York) and SmartQuantum (France). Several other companies also have active research programmes, including Toshiba, HP, IBM, Mitsubishi, NEC and NTT (See External links for direct research links).

Quantum encryption technology provided by the Swiss company Id Quantique was used in the Swiss canton (state) of Geneva to transmit ballot results to the capitol in the national election occurring on Oct. 21, 2007.

In 2004, the world's first bank transfer using quantum cryptography was carried in Vienna. An important cheque, which needed absolute security, was transmitted from the Mayor of the city to an Austrian bank.

The simplest type of possible attack is the intercept-resend attack, where Eve measures the quantum states (photons) sent by Alice and then sends replacement states to Bob, prepared in the state she measures. In the BB84 protocol this will produce errors in the key shared between Alice and Bob. As Eve has no knowledge of the basis a state sent by Alice is encoded in, she can only guess which basis to measure in, in the same way as Bob. If she chooses correctly then she will measure the correct photon polarization state as sent by Alice, and will resend the correct state to Bob. However if she chooses incorrectly then the state she measures will be random, and the state sent to Bob will not be the same as the state sent by Alice. If Bob then measures this state in the same basis Alice sent he will get a random result, as Eve has sent him a state in the opposite basis, instead of the correct result he would get without the presence of Eve.

## Quantum Dot

A quantum dot is a semiconductor whose excitons are confined in all three spatial dimensions. As a result, they have properties that are between those of bulk semiconductors and those of discrete molecules. Researchers have studied quantum dots in transistors, solar cells, LEDs, and diode lasers. They have also investigated quantum dots as agents for medical imaging and hope to use them as qubits. Some quantum dots are commercially available. They contain anywhere from 100 to 1,000 electrons and range from 2 to 10 nanometers, or 10 to 50 atoms, in diameter. At 10 nanometers in diameter, nearly 3 million quantum dots could be lined up end to end and fit within the width of your thumb. These quantum dots confine electrons, holes, or electron-hole pairs or so-called excitons to zero dimensions to a region on the order of the electrons' de Broglie wavelength.

This can be contrasted to quantum wires, which are confined to a line and quantum wells, which are confined to a planar region. This confinement leads to discrete quantized energy levels and to the quantization of charge in units of the elementary electric charge *e*. Quantum dots are particularly significant for optical applications due to their theoretically high quantum yield. Quantum dots have also been suggested as implementations of a qubit for quantum information processing. Because the quantum dot has discrete energy levels, much like an atom, they are sometimes called "artificial atoms". The energy levels can be controlled by changing the size and shape of the quantum dot, and the depth of the potential.

Like in atoms, the energy levels of small quantum dots can be probed by optical spectroscopy techniques. In contrast to atoms it is relatively easy to connect quantum dots by tunnel barriers to conducting leads, which allows the application of the techniques of tunneling spectroscopy for their investigation. One of the optical features of small excitonic quantum dots immediately noticeable to the unaided eye is coloration. While the material which makes up a quantum dot is significant, more significant in terms of coloration is the size.

The larger the dot, the redder (the more towards the red end of the spectrum) the fluorescence. The smaller the dot, the bluer (the more towards the blue end) it is. The coloration is directly related to the energy levels of the quantum dot. Quantitatively speaking, the bandgap energy that determines the energy (and hence color) of the fluoresced light is inversely proportional to the square of the size of the quantum dot. Larger quantum dots have more energy levels which are more closely spaced. This allows the quantum dot to absorb photons containing less energy, i.e. those closer to the red end of the spectrum. Recent articles in Nanotechnology and other journals have begun to suggest that the shape of the quantum dot may well also be a factor in the colorization, but as yet not enough information has become available.

The ability to tune the size

of quantum dots is advantageous, as the larger and more red-shifted the quantum dots, the less the quantum properties are. The small size of the quantum dot allows people to take advantage of these quantum properties.

In semiconductors, quantum dots are small regions of one material buried in another with a larger band gap. Quantum dots sometimes occur spontaneously in quantum well structures due to monolayer fluctuations in the well's thickness. Self-assembled quantum dots nucleate spontaneously under certain conditions during molecular beam epitaxy(MBE) and metallorganic vapor phase epitaxy (MOVPE), when a material is grown on a substrate to which it is not lattice matched. The resulting strain produces coherently strained islands on top of a two-dimensional "wetting-layer". This growth mode is known as Stranski-Krastanov growth. The islands can be subsequently buried to form the quantum dot. This fabrication method has the most potential for applications in quantum cryptography (i.e. single photon sources) and quantum computation. The main limitations of this method are the cost of fabrication and the lack of control over positioning of individual dots.

Individual quantum dots can be created by a technique called electron beam lithography, in which a pattern is etched onto a semiconductor chip, and conducting metal is then deposited onto the pattern.

In large numbers, quantum dots may also be synthesized by means of a colloidal synthesis. Epitaxy, lithography, and colloidal synthesis all have different positive and negative aspects. By far the cheapest, colloidal synthesis also has the advantage of being able to occur at benchtop conditions and is acknowledged to be the least toxic of all the different forms of synthesis.

Highly ordered arrays of quantum dots may also be self assembled by electrochemical techniques. A template is created by causing an ionic reaction at an electrolyte-metal interface which results in the spontaneous assembly of nanostructures, including quantum dots, on the metal which is then used as a mask for mesa-etching these nanostructures on a chosen substrate.

Yet another method is pyrolytic synthesis, which produces large numbers of quantum dots that self-assemble into preferential crystal sizes.

Being quasi-zero dimensional, quantum dots have a sharper density of states than higher-dimensional structures. As a result, they have superior transport and optical properties, and are being researched for use in diode lasers, amplifiers, and biological sensors.

Quantum dots have quickly found their way into homes in many electronics. The new PlayStation 3 and high-definition DVD players (notably Bluray and HD-DVD) to come out all use a blue laser for data reading. The blue laser up until only a few years ago was beginning to be seen as something of an impossibility, until the synthesis of a blue quantum dot laser.

Quantum dots are one of the most hopeful candidates for solid-state quantum computation. By applying small voltages to the leads, one can control the flow of electrons through the quantum dot and thereby make precise measurements of the spin and other properties therein.

With several entangled quantum dots, or qubits, plus a way of performing operations, quantum calculations might be possible.

Another cutting edge application of quantum dots is also being researched as potential artificial fluorophore for intraoperative detection of tumors using fluorescence spectroscopy.

In modern biological analysis, various kinds of organic dyes are used. However, with each passing year, more flexibility is being required of these dyes, and the traditional dyes are simply unable to meet the necessary standards at times. To this end, Quantum Dots have quickly filled in the role, being found to be superior to traditional organic dyes on several counts, one of the most immediately obvious being brightness (owing to the high quantum yield) as well as their stability. Currently under research as well is tuning of the toxicity.

In a paper published in the May 2004 issue of Physical Review Letters a team from Los Alamos National Laboratory found that quantum dots produce as many as three electrons from one high energy photon of sunlight. When today's photovoltaic solar cells absorb a photon of sunlight, the energy gets converted to at most one electron, and the rest is lost as heat. This could boost the efficiency of panels produced in research labs from today's 20-30% to 42%. This work was reproduced one year later by an NREL team.

Another paper, published in the October 18, 2005 issue of the *Journal of the American Chemical Society*, reports that Michael Bowers II at Vanderbilt University discovered that certain size crystals of cadmium and selenium emit white light when excited by an ultraviolet laser. This emission appears to be coming from the surface of the crystal, rather than the center. The crystals contain either 33 or 34 pairs of atoms. While they are being pyrolytically synthesized, they preferentially form into just this size; so Bowers can make a batch of such crystals in about an hour. Another student then mixed these quantum dots into ordinary varnish, applied it to a blue LED, and observed that the emission is yellowish-white, like a light bulb. The researchers believe that it will be possible to achieve this emission of white light via electrical stimulation as well as photonic, and hope to demonstrate it soon. There are several inquiries into using quantum dots to make displays and light sources: "QD-LED" displays, and "QD-WLED" (White LED). In June, 2006, QD Vision announced technical success in making a proof of concept quantum dot display.

Quantum dots are valued for displays, because they are very small, they emit colored light in very specific frequencies, and because they require very little power, since they are entirely self-illuminating.

## Quantum Dot Nanocrystals

Quantum Dot Nanocrystals is Used to tag biological molecules, and "measuring between five and ten nanometres across, are made up of three components. Their cores contain paired clusters of atoms such as cadmium and selenium that combine to create a semiconductor. This releases light of a specific colour when stimulated by ultraviolet of a wide range of frequencies.

## Quantum Mechanics

A largely computational physical theory explaining the behavior of quantum phenomena, which incorporates the theory of special relativity. Despite dilignet attempts, general relativity has not been sucessfully incorporated into quantum mechanics. A physical model of chemical and optical phenomena, as well as the behaviour of matter in general on a small scale. Quantum mechanics describes a system of particles in terms of a wave function defined over the configuration space of the system.

Although the concept of particles having distinct locations is implicit in the potential energy function that determines the wave function (e.g., of a ground-state system), the observable dynamics of the system cannot be described in terms of the motion of such particles from point to point. In describing the energies, distributions, and behaviors of electrons in nanometer-scale structures, quantum mechanical methods are necessary. Electron wave functions help determine the potential energy surface of a molecular system, which in turn is the basis for classical descriptions of molecular motion. Nanomechanical systems can almost always be described in terms of classical mechanics, with occasional quantum mechanical corrections applied within the framework of a classical model.

## Quantum Mirage

A nanoscale property that may allow information to be transfered through use of the wave property of electrons. Thus, quantum computers might not require wires as we know them. A nanoscale property that may allow information to be transfered through use of the wave property of electrons.

## Quantum Physics

Quantum physics is a branch of science that deals with discrete, indivisible units of energy called quanta as described by the Quantum Theory. Quantum electronics, quantum optics and optoelectronics are important areas of application—atomic clocks, lasers, light emitting diodes, optical fibers, tunnel diodes and superconducting systems are important and well-known examples. The trend towards faster and more complicated microprocessors and microelectronics has resulted in electronic components now approaching the domains of quantum physics at research level.

In about 20 years time, miniaturization will also be halted in commercial applications. It will then probably utilize electronic wave phenomena and single- electron effects in semiconductor components and systems and also to create more complicated transistor components connected in more complicated ways where quantum phenomena, cooperative phenomena and even superconductivity may be important for function.

## Quantum Teleportation

'Teleportation' (or copying) of the quantum state from one particle to another. Not really spectacular as in Star Trek serial where macroscopic objects (humans, weapons and similar) are 'teleported'. Present technology can 'teleport' quantum objects such as photons, electrons ans similar. It is important to note that the object itself is not teleported during this process. What actually happens is that the quantum state (e.g. spin number) is exactly copied from one object to another, both spatially and temporally present *prior* to 'teleportation'.

The technique used is based on the Einstein-Podolsky-Rosen effect (EPR) or paradox, which relies on the so-called nonlocality of the wavefunction. An EPR source produces pair of entangled particles, one of which serves as a 'carbon copy'. The third particle (original, source particle to be copied, introduced besides the EPR pair) is 'entangled' with one of the particles from the EPR pair which enables the copying of its state to another EPR particle that has never interacted with the original, yet it becomes its copy. The experimental setup requires one measuring apparatus that entangles the original with one of the EPR particles and another apparatus that acts on another EPR particle which becomes the copy. The experimentalists operating the two apparatuses must be in contact i.e. the one that acts on the copy particle must know the results of the entanglement measurement on the source (original) particle and another EPR particle which is 'waisted' in the process. It should also be noted here that none of the experimentalists does not know the quantum state of the original particle, even when the whole process finishes, neither is it measured during the process.

## Quantum Wire

In condensed matter physics, a quantum wire is an electrically conducting wire, in which quantum effects are affecting transport properties. Due to the confinement of conduction electrons in the transverse direction of the wire, their transverse energy is quantized into a serie of discrete values $E_0$ ("ground state" energy, with lower value), $E_1$.

Instead an exact calculation of the transverse energies of the confined electrons has to be performed to calculate the wire resistance. Following from the quantization of electron energy, the resistance is also found to be quantized.

The importance of the quantization is inversely proportional to the diameter of the nanowire for a given material. From material to material, it is dependent on the electronic properties, especially on the effective mass of the electrons. In simple words, it means that it will depend on how conduction electrons interact with the atoms within a given material. In practice, semiconductors show clear conductance quantization for large wire transverse dimensions (100 nm) because the electronic modes due to confinement are spatially extended. As a result their fermi wavelengths are large and thus they have low energy separations. This means that they can only be resolved at cryogenic temperature (few kelvins) where the thermal excitation energy is lower than the intermode energy separation.

For metals, quantization corresponding to the lowest energy states is only observed for atomic wires. Their corresponding wavelength being thus extremely small they have a very large energy separation which makes resistance quantization perfectly observable at room temperature

It is possible to make quantum wires out of metallic carbon nanotubes, at least in limited quantities.

The advantages of making wires from carbon nanotubes include their high electrical conductivity – (due to a high mobility – light weight, small scale and tensile strength. The major drawback (as of 2005) is cost.

It has been claimed that it is possible to create macroscopic quantum wires. With a rope of carbon nanotubes, it is not necessary for any single fiber to travel the entire length, since quantum tunneling will allow electrons to jump from strand to strand. This makes quantum wires interesting for commercial uses.

In April 2005, NASA invested $11 million over four years with Rice University to develop quantum wire with 10 times better conductivity than copper at one-sixth the weight. It would be made with carbon nanotubes and would help reduce the weight of the next generation shuttle; but can have wide ranging applications.

# R

## Radio Telescope

The first radio antenna used to identify an astronomical radio source was one built by Karl Guthe Jansky, an engineer with Bell Telephone Laboratories, in 1931. Jansky was assigned the job of identifying sources of static that might interfere with radio telephone service. Jansky's antenna was designed to receive short wave radio signals at a frequency of 20.5 MHz (wavelength about 14.6 m). It was mounted on a turntable that allowed it to rotate in any direction, earning it the name "*Jansky's merry-go-round*". It had a diameter of approximately 100 ft (30 m). and stood 20 ft (6 m). tall. By rotating the antenna on a set of four Ford Model-T tires, the direction of the received interfering radio source (static) could be pinpointed. A small shed to the side of the antenna housed an analog pen-and-paper recording system. After recording signals from all directions for several months, Jansky eventually categorized them into three types of static: nearby thunderstorms, distant thunderstorms, and a faint steady hiss of unknown origin. Jansky finally determined that the "faint hiss" repeated on a cycle of 23 hours and 56 minutes. This four-minute lag is typical of an astronomical sidereal day, the time it takes any "fixed" object located on the celestial sphere to come back to the same location in the sky. By comparing his observations with optical astronomical maps, Jansky concluded that the radiation was coming from the Milky Way and was strongest in the direction of the center of the galaxy, in the constellation of Sagittarius.

Grote Reber was one of the pioneers of what became known as radio astronomy when he built the first parabolic "dish" radio telescope (9 m in diameter) in 1937. He was instrumental in repeating Karl Guthe Jansky's pioneering but somewhat simple work, and went on to conduct the first sky survey in the radio frequencies. After World War II, substantial improvements in radio astronomy technology were made by astronomers in Europe, Australia and the United States, and the field of radio astronomy began to blossom.

The range of frequencies in the electromagnetic spectrum that makes up the radio spectrum is very large. This means the variety and types of antennas that are used as radio telescopes vary in design, size, and configuration. At wavelengths of 30 meters to 3 meters (10 MHz - 100 MHz), they are generally directional antenna arrays similar to "TV antennas" or large stationary reflectors with moveable focal points. Since the wavelengths being observed with these types of antennas are so long, the "reflector" surfaces can be constructed from coarse wire mesh. At shorter wavelengths "dish" style radio telescopes predominate. The angular resolution of a dish style antenna is a function of the diameter of the dish in proportion to the wavelength of the electromagnetic radiation being observed. This dictates the size of the dish a radio telescope needs in order to have a useful resolution. Radio telescopes operating at wavelengths of 3 meters to 30 cm (100 MHz to 1 GHz) are usually well over 100 meters in diameter. Telescopes working at wavelengths above 30 cm (1 GHz) range in size from 3 to 90 meters in diameter.

*Big dishes:* In the late 1950s and early 1960s saw the development of large single-dish radio telescopes. The largest individual radio telescope is the RATAN-600 (built in 1977 in the USSR, belongs to Russia since 1991) with 576 meter diameter of circular antenna (RATAN-600 description). Other two individual radio telescopes at Pushchino Radio Astronomy Observatory, Russia, designed specially for the low frequency observations, are between the largest in their class. LPA (LPA description (in Russian)) is 187 x 384 m size phased array meridional radio telescope, and DKR-1000 is 1000 x 1000 m cross radio telescope (DKR-1000 description (in Russian) ). The largest radio telescope in Europe is the 100 meter diameter antenna in Effelsberg, Germany, which also was the largest fully steerable telecope for 30 years until the Green Bank Telescope was opened in 2000. The largest radio telescope in the United States until 1998 was Ohio State University's The Big Ear.

Other well known disk radio telescopes include the Arecibo radio telescope located in Arecibo, Puerto Rico, which is steerable within about 20° of the zenith and is the largest single-aperture telescope (cf. multiple aperture telescope) ever to be constructed, and the fully steerable Lovell telescope at Jodrell Bank in the United Kingdom. A typical size of the single antenna of a radio telescope is 25 metre, dozens of radio telescopes with comparable sizes are operated in radio observatories all over the world.

*Radio interferometry:* One of the most notable developments came in 1946 with the introduction of the technique called astronomical interferometry. Astronomical radio interferometers usually consist either of arrays of parabolic dishes (e.g. the One-Mile Telescope), arrays of one-dimensional antennas (e.g. the Molonglo Observatory Synthesis Telescope) or two-dimensional arrays of omni-directional dipoles (e.g. Tony Hewish's Pulsar Array). All of the telescopes in the array are widely separated and are connected together using coaxial cable, waveguide, optical fiber, or other type of transmission line. This not only increases the total signal collected, it can also be used in a process called Aperture synthesis to vastly increase resolution. This technique works by superposing (interfering) the signal waves from the different telescopes on the principle that waves that coincide with the same phase will add to each other while two waves that have opposite phases will cancel each other out. This creates a combined telescope that is the size of the antennas furthest apart in the array. In order to produce a high quality image, a large number of different separations between different telescopes are required (the projected separation between any two telescopes as seen from the radio source is called a baseline) - as many different baselines as possible are required in order to get a good quality image (For example the Very Large Array (VLA) in Socorro, New Mexico has 27 telescopes giving 351 independent baselines at once to achieve resolution of 0.2 arc seconds at 3 cm wavelengths). Martin Ryle's group in Cambridge obtained a Nobel Prize for interferometry and aperture synthesis. The Lloyd's mirror interferometer was also developed independently in 1946 by Joseph Pawsey's group at the University of Sydney. In the early 1950s the Cambridge Interferometer mapped the radio sky to produce the famous 2C and 3C surveys of radio sources. The largest existing radio telescope array is the Giant Metrewave Radio Telescope, located in Pune, India. A larger array, LOFAR (the 'LOw Frequency ARray') is currently being constructed in western Europe, consisting of 25 000 small antennas over an area several hundreds of kilometres in diameter.

## Radioactive Decay

Radioactivity can occur when the radius of a nucleus is large compared with the radius of the strong force, which only acts over distances on the order of 1 fm. The most common forms of radioactive decay are: (i) Alpha decay is caused when the nucleus emits an alpha particle, which is a helium nucleus consisting of two protons and two neutrons. The result of the emission is a new element with a lower atomic number. (ii) Beta decay is regulated by the weak force, and results from a transformation of a neutron into a proton, or a proton into a neutron. The first is accompanied by the emission of an electron and an antineutrino, while the second causes the emission of a positron and a neutrino. The electron or positron emissions are called beta particles. Beta decay either increases or decreases the atomic number of the nucleus by one. (iii) Gamma decay results from a change in the energy level of the nucleus to a lower state, resulting in the emission of electromagnetic radiation. This can occur following the emission of an alpha or a beta particle from radioactive decay. Other more rare types of radioactive decay include ejection of neutrons or protons or clusters of nucleons from a nucleus, or more than one beta particle, or result (through internal conversion) in production of high-speed electrons which are not beta rays, and high-energy photons which are not gamma rays. Each radioactive isotope has a characteristic decay time period—the half-life—that is determined by the amount of time needed for half of a sample to decay. This is an exponential decay process that steadily decreases the proportion of the remaining isotope by 50% every half life. Hence after two half-lives have passed only 25% of the isotope will be present, and so forth.

### Radiometer

A radiometer is a device for measuring the radiant flux (power) of electromagnetic radiation. Generally, the term "radiometer" denotes an infrared radiation detector, yet it also comprises detectors operating on any electromagnetic wavelength, e.g. spectroradiometer.

### Rational Drug Design

The input of biocomputing in drug discovery is twofold: firstly the computer may help to optimise the pharmacological profile of existing drugs by guiding the synthesis of new and "better" compounds. Secondly, as more and more structural information on possible protein targets and their biochemical role in the cell becomes available, completely new therapeutic concepts can be developed. The computer helps in both steps: to find out about possible biological functions of a protein by comparing its amino acid sequence to databases of proteins with known function, and to understand the molecular workings of a given protein structure. Understanding the biological or biochemical mechanism of a disease then often suggests the types of molecules needed for new drugs

### Reactant

A substance that is consumed

in the course of a chemical reaction. It is sometimes known, especially in the older literature, as a reagent, but this term is better used in a more specialized sense as a test substance that is added to a system in order to bring about a reaction or to see whether a reaction occurs (e.g. an analytical reagent).

## Reagent Device

A large reagent structure (or a large structure that binds a smaller reagent) serving as a component of a mechanochemical system. A reagent device exists chiefly to hold, position, and manipulate the environment of a reagent moiety.

## Recessive

Refers to the member of a pair of alleles that fails to be expressed in the phenotype of the organism when the dominant member is present. Also refers to the phenotype of an individual that has only the recessive allele.

## Recombinant DNA

Any DNA molecule formed by joining DNA segments from different sources. Recombinant DNAs are widely used in the cloning of genes, in the genetic modification of organisms, and in molecular biology generally.

## Recombination

Process in which chromosomes or DNA molecules are broken and the fragments are rejoined in new combinations. Can occur in the living cell—for example, through crossing-over during meiosis—or in the test tube using purified DNA and enzymes that break and ligate DNA strands.

## Reconnaisance Satellite

Also called a spy satellite because it is used to spy on other countries. It can provide intelligence information on military activities, detect missile launches or nuclear explosions, and pick up and record radio and radar transmissions while passing over a country. It can also be used as an orbital weapon by placing warheads on a low orbit satellite to be launched at a ground target, but this is not a recommended or frequent use of satellites.

## Reconstruction

A crystal consists of a regular array of atoms, and the simplest model of a crystal surface would be generated by simply discarding all atoms to one side of a surface without changing the positions of the rest. In reality, however, the positions of the remaining atoms do change. A pattern of displacements that lowers the symmetry of the surface (relative to the ideally terminated crystal) is termed a surface reconstruction; some reconstructions alter the pattern of bonds.

### Reduced Mass

Many dynamical properties of a system consisting of two interacting masses, $m_1$ and $m_2$, are equivalent to those of a system in which one mass is fixed in space and the other has a mass (the reduced mass) with the value $m_1m_2/(m_1 + m_2)$. The reduced mass description has fewer dynamical variables.

### Relaxation Time

A measure of the rate at which a disequilibrium distribution decays toward an equilibrium distribution. The electron relaxation time in a metal, for example, describes the time required for a disequilibrium distribution of electron momenta (e.g., in a flowing current) to decay toward equilibrium in the absence of an ongoing driving force and can be interpreted as the mean time between scattering events for a given electron.

### Remote Sensing Satellite

A type of satellite that performs remote sensing from space. Remote sensing satellites generally monitor important resources for humans. For example, they might be used to track animal migration, locate mineral deposits, watch agricultural crops for weather damage, or see how fast the forests are being cut down. Because they are in space, remote sensing satellites are ideal for monitoring areas with harsh climates or difficult terrains.

### Remote Sensing

The process of collecting data about something from a point far away. A satellite making observations of the Earth, therefore, is using its equipment for remote sensing. The mechanism used in living systems to make copies of genetic information.

### Replicator

A system able to build copies of itself when provided with raw materials and energy. Any system that can build copies of itself when provided with the appropriate raw materials and energy. In discussions of evolution, a replicator is an entity (such as a gene, a meme, or the contents of a computer memory disk) which can get itself copied, including any changes it may have undergone. In a broader sense, a replicator is a system which can make a copy of itself, not necessarily copying any changes it may have undergone. A rabbit's genes are replicators in the first sense (a change in a gene can be inherited); the rabbit itself is a replicator only in the second sense (a notch made in its ear can't be inherited).

### Reporter Gene

Reporter Gene detection has become invaluable in the study of gene function and related cellular events. In these studies, the reporter gene acts as a surrogate for the coding region of the gene under study. Upon transfection of the reporter construct, expression of the reporter gene may be monitored by direct (e.g. enzymatic) or indirect (e.g. immunochemical) assay of the reporter protein. Localized expression of the reporter protein can be detected via immunochemistry using antibodies specific for the expressed reporter protein.

### Reproducibility

A property of an experiment or process where one tends to receive consistent results from following a specific procedure.

### Respirocyte

Respirocytes are hypothetical, artificial red blood cells that can supplement or replace the function of much of the human body's normal respiratory system. Still entirely theoretical, respirocytes measure 1 micrometer in diameter. Respirocytes mimic the action of the natural hemoglobin-filled red blood cells. The design of the spherical nanorobot is made up of 18 billion atoms arranged as a tiny pressure tank. The tank can be filled up with oxygen and carbon dioxide, making one complete transfer point at the lungs, and the reverse transfer at the body's tissues.

The uses of each respirocyte can store and transport 200 times more oxygen and carbon dioxide than natural red blood cells. Filled with these respirocytes, an adult human could hold his/her breath underwater for four hours. That person could also sprint at top speed for at least 15 minutes without taking a breath.

Muscle fatigue results from inadequate supply of oxygen to the muscles during intense exercise, leading to inefficient anaerobic respiration and the buildup of lactic acid in muscle tissue, causing pain and exhaustion. If respirocytes could increase the supply of oxygen despite exercise, it should be possible to reduce muscle fatigue, increasing a person's endurance. Theorist Robert Freitas has proposed respirocytes as a superior alternative to naturally occurring red blood cells, and has similarly proposed "microbivore" robots that would attack pathogens in the manner of white blood cells.

By definition, respirocytes qualify as nanotechnology, a field of technology still in the early phases of development. Considerations involved in building respirocytes include power, immune reaction or toxicity, computation and communication, and social issues such as economics. Because respirocytes and related technologies would, if successful, improve the user's abilities beyond normal human limits, their design is associated with the Transhumanism movement which seeks such advances.

## Robot

A robot is a mechanical or virtual artificial agent. In practice, it is usually an electro-mechanical system which, by its appearance or movements, conveys a sense that it has intent or agency of its own. The word *robot* can refer to both physical robots and virtual software agents, but the latter are usually referred to as bots. There is no consensus on which machines qualify as robots, but there is general agreement among experts and the public that robots tend to do some or all of the following: move around, operate a mechanical arm, sense and manipulate their environment, and exhibit intelligent behavior, especially behavior which mimics humans or animals.

Stories of artificial helpers and companions and attempts to create them have a long history, but fully autonomous machines only appeared in the 20th century. The first digitally operated and programmable robot, the Unimate, was installed in 1961 to lift hot pieces of metal from a die casting machine and stack them.

Today, commercial and industrial robots are in widespread use performing jobs more cheaply or with greater accuracy and reliability than humans. They are also employed for jobs which are too dirty, dangerous or dull to be suitable for humans. Robots are widely used in manufacturing, assembly and packing, transport, earth and space exploration, surgery, weaponry, laboratory research, and mass production of consumer and industrial goods. People have a generally positive perception of the robots they actually encounter. Robotic competitions are popular, and provide training as well as entertainment for technically-inclined students. Domestic robots for cleaning and maintenance and robotic toys are increasingly common in and around homes. Asians and Westerners have different expectations for the future of consumer robotics, but these expectations are generally positive.

There is anxiety, however, over the economic impact of automation and the threat of robotic weaponry, anxiety which is not helped by the many villainous, intelligent, acrobatic robots in popular entertainment. Compared with their fictional counterparts, real robots are still benign, dim-witted and clumsy.

# S

## Satellite DNA

Satellite DNA consists of highly repetitive DNA, and is so called because repetitions of a short DNA sequence tend to produce a different frequency of the nucleotides adenine, cytosine, guanine and thymine, and thus have a different density from bulk DNA - such that they form a second or 'satellite' band when genomic DNA is separated on a density gradient.

A repeated pattern can be between 1 base pair long (a mononucleotide repeat) to several thousand base pairs long, and the total size of a satellite DNA block can be several megabases without interruption. Most satellite DNA is localized to the telomeric or the centromeric region of the chromosome. The nucleotide sequence of the repeats is fairly well conserved across a species. However, variation in the length of the repeat is common. For example, minisatellite DNA is a short region (1-5kb) of 20-50 repeats. The difference in how many of the repeats is present in the region (length of the region) is the basis for DNA fingerprinting.

## Scanning Force Microscope

Scanning Force Microscope invented in 1986 by Binnig et al., scanning force microscopy has become a powerful tool for the investigation of surfaces. A great advantage of this method is the high resolution—down to the atomic scale—which can be achieved even in air and even on insulating surfaces. However, the greatest advantage is the possibility to investigate surfaces *in situ* in a liquid phase.

## Scanning Near Field Optical Microscopy

A method for observing local optical properties of a surface that can be smaller than the wavelength of the light used. A type of scanning probe microscopy that maps the near-field optical properties of an interface.

## Schwann Cell

Named after the German physiologist Theodor Schwann, Schwann cells (also referred to as neurolemmocytes) are a variety of glial cell that mainly provide myelin insulation to axons in the peripheral nervous system of jawed vertebrates. The vertebrate nervous system relies on this myelin sheath for insulation and as a method of decreasing membrane capacitance in the axon, thus allowing for saltatory conduction to occur and for an increase in impulse speed, without an increase in axonal diameter. Non-myelinating Schwann cells are involved in maintenance of axons and are crucial for neuronal survival. Some group around smaller axons and form Remak bundles. Schwann cells are the peripheral nervous system's analogues of the central nervous system oligodendrocytes.

Schwann cells begin to form the myelin sheath in mammals during fetal development and work by spiraling around the axon, sometimes with as many as 100 revolutions. A well-developed Schwann cell is shaped like a rolled-up sheet of paper, with layers of myelin in between each coil. The inner layers of the wrapping, which are predominantly membrane material, form the myelin sheath while the outermost layer of nucleated cytoplasm forms the neurolemma. Only a small volume of residual cytoplasm communicates the inner from the outer layers. This is seen histologically as the Schmidt-Lantermann Incisure. Since each Schwann cell can cover about a millimeter (0.04 inches) along the axon, hundreds and often thousands are needed to completely cover an axon, which can sometimes span the length of a body. The gaps between the Schwann cell covered segments are the Nodes of Ranvier, important sites of ionic and other exchanges of the axon with the extracellular liquid.

Unlike oligodendrocytes, myelinating Schwann cells provide insulation to only one axon (see image). This arrangement permits saltatory conduction of action potentials which greatly speeds it and saves energy.

A number of experimental studies since 2001 have implanted Schwann cells in an attempt to induce remyelination in multiple sclerosis-afflicted patients.The Schwann cells are the cells that make the myelin in the peripheral nervous system (PNS) Indeed, Schwann cells are known for their roles in supporting nerve regeneration.

## Scintillation

Burst of luminescence of short duration caused by an individual energetic particle.

## Sealed Assembler Lab

A general-purpose assembler system in a container permitting only energy and information to be exchanged with the environment.

## Sealed Assembler Laboratory

A work space, containing assemblers, encapsulated in a way that allows information to flow in and out but does not allow the escape of assemblers or their products.

## Secondary Structure

Regular local folding pattern of a polymeric molecule; in proteins, a helices and b-pleated sheets.

## Secretory Vesicle

Membrane-bounded organelle in which molecules destined for secretion are stored prior to release. Sometimes called secretory granule because darkly staining contents make the organelle visible as a small solid object.

## Selectivity

The ability of a sensor to measure only one metric or, in the case of a chemical sensor, to measure only a single chemical species.

## Self-Assembled Monolayer

A two-dimensional film, one molecule thick, covalently assembled at an interface.

## Self-Assembler

A specific type of assembler that makes use of self-assembly such that the assembly process would theoretically not require external energy or information input.

### Self-assembling Biomolecular Materials

Examples of self- assembly include protein folding, the formation of liposomes, and the alignment of liquid crystals. While this type of equilibrium self- assembly is the central focus of this report, it is important to emphasize that much biological assembly is also driven by energy sources such as adenosine triphosphate (ATP), which power biomotors.

### Self-replication

More accurately labeled "exponential replication," self-replication refers to the process of growth or replication involving doubling within a given period. Example: create one assembler. Program it to create another, and program that one likewise, etcetera, until you have a speficied amount. Self replication is an effective route to truly low cost manufacturing. Our intuitions about self replicating systems, learned from the biological systems that surround us, are likely to seriously mislead us about the properties and characteristics of artificial self replicating systems designed for manufacturing purposes. Artificial systems able to make a wide range of non-biological products (like diamond) under programmatic control are likely to be more brittle and less adaptable in their response to changes in their environment than biological systems. At the same time, they should be simpler and easier to design. The complexity of such systems need not be excessive by present engineering standards. The act of objects making copies of themselves—similar to biological reproduction except that self-replicating objects should make exact copies of themselves.

### Sensitivity

The amount of change in a sensor's output in response to a change at a sensor's input over the sensor's entire range. Provides an indication of a sensor's ability to detect changes. For some sensors, the sensitivity is defined as the input parameter change required to produce a standardized output change.

## Sensors

A sensor is a device that measures a physical quantity and converts it into a signal which can be read by an observer or by an instrument. For example, a mercury thermometer converts the measured temperature into expansion and contraction of a liquid which can be read on a calibrated glass tube. A thermocouple converts temperature to an output voltage which can be read by a voltmeter. For accuracy, all sensors need to be calibrated against known standards.

Sensors are used in everyday objects such as touch-sensitive elevator buttons and lamps which dim or brighten by touching the base. There are also innumerable applications for sensors of which most people are never aware. Applications include cars, machines, aerospace, medicine, manufacturing and robotics.

A sensor's sensitivity indicates how much the sensor's output changes when the measured quantity changes. For instance, if the mercury in a thermometer moves 1 cm when the temperature changes by 1 °C, the sensitivity is 1 cm/°C. Sensors that measure very small changes must have very high sensitivities. Technological progress allows more and more sensors to be manufactured on a microscopic scale as microsensors using MEMS technology. In most cases, a *microsensor* reaches a significantly higher speed and sensitivity compared with macroscopic approaches. See also MEMS sensor generations.

## Sentience Quotient

In the article "Xenopsychology" by Robert Freitas in Analog of April 1984 there is an interesting index called "Sentience quotient". It is based on: The sentience of an intelligence is roughly directly related to the amount of data it can process per unit time and inversely to the overall mass needed to do that processing.

This would be something like baud/kilograms. And since that would rapidly turn into a real big number, base 10 logs are used. The "least sentient" would be one bit over the lifetime of the universe massing the entire known universe, or about -70. The "most sentient" is claimed to be +50. Homo sapiens are around +13, a Cray I is +9, a venus flytrap is a peak of +1 with plants generally -2.

### Shape Memory Alloys

Shape Memory Alloys (SMAs) are a unique class of alloys which are able to "remember" their shape and are able to return to that shape even after being bent. The ability is known as the shape memory effect. ... This property has lead to many uses of SMA from orthodontics and coffee makers to methods of controlling aircraft and protecting buildings from earthquake damage. ... The first SMA to be discovered and the most commonly used is called Nitinol.

### SI Units

International System of units based on the metric system and units derived from the metric system.

### Sigma Bond

A covalent bond in which overlap between two atomic orbitals (e.g., of *sp*, *sp2*, or *sp3* hybridization) produces a single bonding orbital in which the distribution of shared electrons has a roughly cylindrical symmetry about the axis linking the two atoms; see pi bond, single bond, double bond, triple bond. By themselves, sigma bonds present little barrier to rotation of one substructure with respect to another, although steric effects and cyclic structures may hinder or block rotation.

### Signal to Noise

Ratio which can interfere with detection. Can also refer to data analysis. Biological data is often very "noisy". This is particularly seen when trying to look at low abundance biomolecules.

### Single Cell Repair Unit

A cell repair unit using cilia for propulsion and equipped with a nanocomputer having 10 megabytes of fast RAM and 1 gigabyte of slower-access memory. The unit is extending 1000 individually-controlled molecular manipulators.

### Single Crystal

A crystalline solid for which the periodic and repeated atomic pattern extends throughout its entirety without interruption.

## Single Electron Devices

Nanoscale devices that control the movement of individual electrons, may one day make it possible for integrated circuits to have as many as 10 billion electronic devices in a square centimeter, a density 1000 times greater than that believed feasible for conventional integrated circuits. In development since the mid- 1980s, these devices consist of two electrodes (typically 30 nm wide) separated by a 1 nm- deep insulating layer through which single electrons can tunnel. These devices have many potential applications, from building more sensitive measurement devices to understanding fundamental problems in physics. In the last several years, researchers have built two- junction devices that share a middle electrode. These devices are called *"single- electron transistors,"* because, like conventional transistors, their current can be controlled by modifying the surface charge on the middle electrode, making it an ideal element for an integrated circuit. A circuit made of single-electron devices, however, would have to be operated at a temperature of 4 K or below to reduce thermal effects which disturb the movements of single electrons in the solid.

## Single-pair FRET spFRET

Designed to overcome the averaging effects of ensemble studies because measurements are made on single molecules freely diffusing in solution. This method limits the observation period to the diffusion time of each molecule through the focal spot of a laser on the order of a few hundred milliseconds, but it permits the rapid gathering of data at single- molecule resolution on a large number of molecules in a short time period. SpFRET can be used to study intramolecular conformational changes by placing the donor and acceptor fluorescent tags on two different sites of the same macromolecule, or alternatively, intermolecular interactions can be studied by attaching the donor and acceptor tags to two different macromolecules.

## Singularity

Defined by Vernor Vinge as the "postulated point or short period in our future when our self-guided evolutionary development accelerates enormously (powered by nanotechnology, neuroscience, AI, and perhaps uploading) so that nothing beyond that time can reliably be conceived. ...a future time when societal, scientific and economic change is so fast we cannot even imagine what will happen from our present perspective, and when humanity will become posthumanity."

### Sky Hook

A long, very strong, cable in orbit around a planet which rotates around its center of mass in such a way that when one end is closest to the ground, its relative velocity is almost zero. It would function as a kind of space elevator; shuttle craft would anchor to the end and then be lifted into orbit where they would be released. It is closely related to the idea of a beanstalk.

### Slip Casting

A forming technique used to shape ceramic materials. A slip or suspension of solid particles in water is poured into a porous mold. A solid layer forms on the inside wall as water is absorbed by the mold, leaving a shell (or a solid piece) in the shape of the mold.

### Small Molecule Therapeutics

Low molecular-weight drug. Compared to larger molecular weight pharmaceuticals such as proteins, peptides, and carbohydrates, small molecules can more easily penetrate cell membranes and the blood brain barrier. Can be delivered orally or intravenously. These molecules tend to incur lower process development and manufacturing costs.

### Small Molecules

Preferred for drugs as they are orally available (unlike proteins which must be administered by injection or topically). Size of small molecules is generally under 1000 Daltons, but many estimates seem to range between 300 to 700 Daltons.

### Smart Materials and Products

Here, materials and products capable of relatively complex behavior due to the incorporation of nanocomputers and nanomachines. Also used for products having some ability to respond to the environment.

## Smart Sensor

A sensor in which the electronics that process the output from the sensor, and forms the modifier, are partially or fully integrated on a single chip.

## Soft Lithography

In technology, soft lithography refers to a family of techniques for fabricating or replicating structures using "elastomeric stamps, molds, and conformable photomasks" (in the words of Rogers and Nuzzo, p. 50, as cited in "References"). It is called "soft" because it uses elastomeric materials most notably PDMS. Soft lithography is generally used to construct features measured on the micrometer to nanometer scalé. According to Rogers and Nuzzo (2005), development of soft lithography expanded rapidly during the period 1995 to 2005.

## Spectrograph

An instrument that spreads out the light gathered by a telescope so that it can be analyzed to determine many different properties of celestial objects. These properties include the chemical composition and abundance of different elements, temperature, radial velocity, rotational velocity, and magnetic fields.

## Spectrometer

A spectrometer is an optical instrument used to measure properties of light over a specific portion of the electromagnetic spectrum, typically used in spectroscopic analysis to identify materials. The variable measured is most often the light's intensity but could also, for instance, be the polarization state. The independent variable is usually the wavelength of the light, normally expressed as some fraction of a meter, but sometimes expressed as some unit directly proportional to the photon energy, such as wavenumber or electron volts, which has a reciprocal relationship to wavelength. A spectrometer is used in spectroscopy for producing spectral lines and measuring their wavelengths and intensities. Spectrometer is a term that is applied to instruments that operate over a very wide range of wavelengths, from gamma rays and X-rays into the far infrared. If the region of interest is restricted to near the visible spectrum, the study is called spectrophotometry.

In general, any particular instrument will operate over a small portion of this total range because of the different techniques used to measure different portions of the spectrum. Below optical frequencies (that is, at microwave and radio frequencies), the spectrum analyzer is a closely related electronic device.

Spectroscopes are often used in astronomy and some branches of chemistry. Early spectroscopes were simply prisms with graduations marking wavelengths of light. Modern spectroscopes, such as monochromators, generally use a diffraction grating, a movable slit, and some kind of photodetector, all automated and controlled by a computer. The spectroscope was invented by Joseph von Fraunhofer.

When a material is heated to incandescence it emits light that is characteristic of the atomic makeup of the material. Particular light frequencies give rise to sharply defined bands on the scale which can be thought of as fingerprints. For example, the element sodium has a very characteristic double yellow band known as the Sodium D-lines at 588.9950 and 589.5924 nanometers, the colour of which will be familiar to anyone who has seen a low pressure sodium vapor lamp.

In the original spectroscope design in the early 19th century, light entered a slit and a collimating lens transformed the light into a thin beam of parallel rays. The light was then passed through a prism (in hand-held spectroscopes, usually an Amici prism) that refracted the beam into a spectrum because different wavelengths were refracted different amounts due to dispersion. This image was then viewed through a tube with a scale that was transposed upon the spectral image, enabling its direct measurement.

With the development of photographic film, the more accurate spectrograph was created. It was based on the same principle as the spectroscope, but it had a camera in place of the viewing tube. In recent years the electronic circuits built around the photomultiplier tube have replaced the camera, allowing real-time spectrographic analysis with far greater accuracy. Arrays of photosensors are also used in place of film in spectrographic systems. Such spectral analysis, or spectroscopy, has become an important scientific tool for analyzing the composition of unknown material and for studying astronomical phenomena and testing astronomical theories. The wavelengths are measured with the spectrometer.

A spectrograph is an instrument that separates an incoming wave into a frequency spectrum. There are several kinds of machines referred to as *spectrographs*, depending on the precise nature of the waves. The first spectrographs used photographic paper as the detector. The star spectral classification and discovery of the main sequence, Hubble's law and the Hubble sequence were all made with spectrographs that used photographic paper. The plant pigment phytochrome was discovered using a spectrograph that used living plants as the detector. More recent spectrographs use electronic detectors, such as CCDs which can be used for both visible and UV light. The exact choice of detector depends on the wavelengths of light to be recorded.

The forthcoming James Webb Space Telescope will contain both a near-infrared spectrograph (NIRSpec) and a mid-infrared spectrometer (MIRI).

An echelle spectrograph uses two diffraction gratings, rotated 90 degrees with respect to each other and placed close to one another. Therefore an entrance point and not a slit is used and a 2d CCD-chip records the spectrum. Usually one would guess to retrieve a spectrum on the diagonal, but when both grating have a wide spacing and one is blazed so that only the first order is visible and the other is blazed that a lot of higher orders are visible, one gets a very fine spectrum nicely folded onto a small common CCD-chip. The small chip also means that the collimating optics need not to be optimized for coma or astigmatism, but the spherical aberration can be set to zero. A spectrograph is sometimes called polychromator, as an analogy to monochromator.

## Spectroscopic Dyes

These dyes—in particular, nanoparticles — are emerging as alternatives to fluorescent dyes. Because the emission spectra of nanoparticles vary according to these particles' specific size and shape, these nanostructures can be used in multicolor detection formats, potentially offering much greater multiplexing than is currently achievable. The fact that chemists have been able to create a great variety of structures (and properties) in nanoparticles suggests that these particles might be more "finely tunable" than organic dyes, allowing better results from biological assays.

## Spintronics

Spintronics, the technology utilizing the electron spin to manipulate (calculate) or store information. The 'classical' electronics, with respect to spintronics should possibly be called 'chargetronics' since it uses electron charge, i.e. charge currents to manipulate the information (fire the transistors and similar). The spin is the quantum number of an electron which can take two values (states)— 'up' and 'down'. In spintronics, the *spin* currents are of relevance to the function of electronic devices. Note that the spin currents are also the charge currents, since the carrier of spin (electron) is also a carrier of charge. The materials that would support the spintronical applications would have to be sensitive to the sign (up or down) of the spin currents, i.e. their response should be dependent on the sign of the spin current. Materials containing several layers of different magnetic and nonmagnetic metals that have thicknesses in the range of several nanometers are the candidates for such applications. This, of course, requires material nanostructuring.

## SPR Surface Plasmon Resonance

A biosensing technique in which biomolecules capable of binding to specific analytes or ligands are first immobilized on one side of a metallic film. Light is then focused on the opposite side of the film to excite the surface plasmons, that is, the oscillations of free electrons propagating along the film's surface. The refractive index of light reflecting off this surface is measured. When the immobilized biomolecules are bound by their ligands, an alteration in surface plasmons on the opposite side of the film is created which is directly proportional to the change in bound, or adsorbed, mass. Binding is measured by changes in the refractive index. The technique is used to study biomolecular interactions, such as antigen—antibody binding.

### Squeeze-film Damping

Effect of ambient fluid and spacing on the vertical movement of a structural member with respect to a substrate.

### Stable

Strictly speaking, a system is termed stable if no rearrangement of its parts can form a system of lower free energy. In practice, the term is used with an implicit proviso regarding the transformations to be considered. Hydrogen is not considered unstable merely because it is subject to nuclear fusion at extreme temperatures. A system is usually regarded as stable if its rate of transformation to a state of lower free energy is negligible (by some standard) under the ambient conditions. In nanomechanical systems, a structure can commonly be regarded as stable if it has an extremely low rate of transformations when subjected to its intended operating conditions.

### Staining and Labeling

The marking of biological material with a dye or other reagent for the purpose of identifying and quantitating components of tissues, cells or their extracts.

### Statistical Process Control

A method to control product quality by observing a process during its operation while the products are being produced, rather than depending on inspection. This identifies the causes of variation in product quality.

### STEM Antennas

An acronym for Storable Tubular Extendible Member. STEM antennas are stored in the body of a satellite to save space at launch and unrolled when the satellite reaches orbit. Canada's first satellite,

Alouette, used four STEM antennas.

### Step Response

The response of a system to an instantaneous jump in the input signal.

### Steric Hindrance

Slowing of the rate of a chemical reaction owing to the presence of structures on the reagents that mechanically interfere with the motions associated with the reaction, typically by obstructing the reaction site.

### Steric

Pertaining to the spatial relationships of atoms in a molecular structure, and in particular, to the space-filling properties of a molecule. If molecules were rigid and had hard surfaces, steric properties would merely be an opaque way of saying "shape"; a flexible side-chain, however, has definite steric properties but no fixed shape. Nanomechanical systems make extensive use of the steric properties of relatively rigid molecules, for which the term "shape" has essentially its conventional meaning so long as one remembers that the surface interactions are soft on small length-scales.

### Stiffness

The stiffness of a system with respect to a deformation (e.g., the stiffness of a spring with respect to stretching) is the second derivative of the energy with respect to the corresponding displacement; this measures the curvature of the potential energy surface along a particular direction. Positive stiffness is associated with stability, and a large stiffness can result in a small positional uncertainty in the presence of thermal excitation. Negative stiffnesses correspond to unstable locations on the potential energy surface. Alternative terms for stiffness include force gradient and rigidity.

### Stoichiometry

For ionic compounds, the state of having exactly the ratio of cations to anions specified by the chemical formula. Stoichiometric quantities refers to quantities of reactants mixed in exactly the correct amounts so that all are used up at the same time.

### Strain Gauge

An element (wire or foil) that measures a strain based on electrical resistance changes of the gauge that result from a change in length or dimension strain of the wire or foil.

### Strain

In mechanical engineering, strain is a measure of the deformation resulting from stress (that is, force per unit area); the displacement of one point with respect to another, divided by their equilibrium separation in the absence of stress. In chemistry, a molecular fragment generally has some equilibrium geometry (bond lengths, interbond angles, etc.) when the rest of the molecular structure does not impose special constraints (e.g., bending bonds to form a small ring). Deviations from this equilibrium geometry are described as strain, and increase the energy of the molecule. Strain in the mechanical engineering sense causes strain in the chemical sense.

### Structure Activity Relationship (SAR)

The relationship between chemical structure and pharmacological activity for a series of compounds. Compounds are often classed together because they have structural characteristics in common including shape, size, stereochemical arrangement, and distribution of functional groups. Other factors contributing to structure- activity relationship include chemical reactivity, electronic effects, resonance, and inductive effects.

### Structure Based Drug Design

Structure- Based Drug Design and Virtual Screening bring together high- throughput crystallography and NMR, computational advances in docking algorithms and virtual screening, and traditional techniques of high-throughput screening and combinatorial chemistry to increase the efficiency and speed of the drug discovery and development process. ... the latest developments in de novo, NMR-based, and crystallography-based drug design. For years researchers have sought a more rational approach to designing drugs rather than screening for hits and leads. The rapidly growing body of structural information emerging as a result of genomic- derived targets and industrialization of protein structure determination is dramatically altering the data drug designers have to work with. As a result very encouraging progress is being made in drug design. New approaches, such as co-crystallization of ligands with a given target, allows structural techniques to be used for screening, which then further facilitates the design process. Since the early 1980s, industry has been interested in structural biology as part of the discipline of direct structure- based drug design, which combines structural biology with computational and medicinal chemistry in order to design drugs—rather than merely selecting drugs—that modulate a protein target of interest.

## Sun-synchronous Orbit

A sun-synchronous orbit describes the orbit of a satellite that provides consistent lighting of the Earth-scan view. The satellite passes the equator and each latitude at the same time each day. For example, a satellite's sun-synchronous orbit might cross the equator twelve times a day each time at 3:00 p.m. local time. The orbital plane of a sun-synchronous orbit must also precess (rotate) approximately one degree each day, eastward, to keep pace with the Earth's revolution around the sun.

# T

## Temperature Sensitive (ts) Mutant

Organism or cell carrying a genetically altered protein (or RNA molecule) that performs normally at one temperature but is abnormal at another (usually higher) temperature.

## Temperature

A system in which internal vibrational modes have equilibrated with one another can be said to have a particular temperature. Two systems A and B are said to be at different temperatures if, when brought into contact, heat flows from (say) A to B, increasing the thermal energy of B at the expense of the thermal energy of A.

## Template

An external source of information or structure often used in manufacturing and mass production. As a nanoscale example, biological systems are known to use nucleic acids as a templates for protein synthesis.

## Tensile Strength

The maximum engineering stress, in tension, sustainable without fracture; also called"ultimate (tensile) strength".

## Terraform

To change the properties of a planet to make it more earth-like, making it possible for humans or other terrestrial organisms to live unaided on it, for example by changing atmospheric composition, pressure, temperature or the climate and introducing a self-sustaining ecosystem. This will most probably be a very long-term project, probably requiring self-replicating technology and megascale engineering. So far Venus and especially Mars looks as the most promising candidates for terraforming in the solar system.

### Thermal Blanketing

Like insulation in a house, the covering on a satellite that regulates temperature. A material commonly used for thermal blanketing on a satellite is Mylar.

### Thermal Energy

The internal energy present in a system as a result of the energy of thermally equilibrated vibrational modes and other motions (including both kinetic energy and molecular potential energy). The mean thermal energy of a classical harmonic oscillator is kT.

### Thermal Fluctuations (TF)

The thermal energy of a system (or of a particular part or mode of motion in a system) has a mean value determined by the temperature and by the structure of the system. Statistical deviations about that mean are termed TF; these are of great importance in determining both rates of chemical reactions and error rates in nanomechanical systems.

### Thermal Noise

the vibration and motion of atoms and molecules caused by the fact that they have a temperature above absolute zero. Once used as an argument on why MNT could not work.

### Thermal Vacuum Chamber

The closest thing to space on Earth. At the David Florida Laboratory, satellites are tested in this chamber which can simulate the temperature and vacuum of space to measure the cool-down and warm-up characteristics of a piece of hardware.

### Thermistor

A thermistor is a type of resistor with resistance varying according to its temperature. The word is a portmanteau of *thermal* and *resistor*. Samuel Ruben invented the thermistor in 1930, and was awarded U.S. Patent No. 2,021,491.

Thermistors are widely used as inrush current limiters, temperature sensors, self-resetting overcurrent protectors, and self-regulating heating elements. Thermistors can be classified into two types depending on the sign of $k$. If $k$ is positive, the resistance increases with increasing temperature, and the device is called a positive temperature coefficient (PTC) thermistor, or posistor. If $k$ is negative, the resistance decreases with increasing temperature, and the device is called a negative temperature coefficient (NTC) thermistor. Resistors that are not thermistors are designed to have a $k$ as close to zero as possible, so that their resistance remains nearly constant over a wide temperature range.

Thermistors differ from resistance temperature detectors (RTD) in that the material used in a thermistor is generally a ceramic or polymer, while RTDs use pure metals. The temperature response is also different; RTDs are useful over larger temperature ranges, while thermistors typically achieve a higher precision within a limited temperature range.

## Thermocouple

A temperature-measuring device, which contains a pair of end-joined dissimilar conductors in which an electromotive force is developed by thermoelectric effects when the joined ends and the free ends of the conductors are a different temp.

## Thermocycler

An instrument that repeatedly cycles through various temperatures required for an iterative, temperature-dependant chemical process such as the polymerase chain reaction.

## Thermodynamical Stability

Stability of a certain system with respect to changes in thermodynamical parameters. For nanoscopic systems, the most important parameter is temperature, i.e. stability of a system with respect to increase of temperature. Some of nanoscopic structures are stable only at sufficiently low temperatures, while at higher temperatures their structure tends to degrade.

At higher temperatures, the atoms of which a nanostructure is built vibrate with larger amplitudes which may result in degradation of the structure. Additionally, nanostructured material, or a nanoparticle may be unstable when exposed to chemical influences from the surrounding, i.e. atoms and molecules. Some nanostructures are stable only in ultra-high vacuum conditions, which very much limits their use. Thermodynamical and chemical stability is a major concern in many applications involving nanoparticles and nanostructured materials.

### Thermoelastic

Both stress and temperature changes alter the dimensions of an object having a finite stiffness and a nonzero thermal expansion coefficient. Applying a stress then produces a temperature change; this can result in a heat flow which then changes the stress: these are thermoelastic effects, and result in losses of free energy.

### Thermogenic Limit

In medical nanorobotics, the maximum amount of waste heat that may safely be released by a population of in vivo medical nanorobots that are operating within a given tissue volume.

### Thermoplastic Polymer

A substance that when molded to a certain shape under appropriate conditions can later be remelted.

### Thioester Bond

High-energy bond formed by a condenzation reaction between an acid (acyl) group and a thiol group (–SH); seen, for example, in acetyl CoA and in many enzyme-substrate complexes.

### Thiol

An SH group, or a molecule containing one. Also known as a sulfhydryl or mercapto group.

### Three-axis Stabilized

Done with an internal gyroscope and thrusters. A gyroscope's stable spin can be used as a sensor to tell the satellite when its attitude is changing. The satellite can then correct the problem using thrusters. This is one form of attitude control.

### Thruster

A way of controlling a satellite's attitude. Thrusters usually contain compressed gas that when sent out of the end of the thruster will move the satellite in space. The force of the compressed gas (the action) causes the satellite to move in the opposite direction (the reaction). This is Newton's third law of motion—for every action there is an equal and opposite reaction.

## Thylakoid

Flattened sac of membrane in a chloroplast that contains pigment and carries out the light-gathering reactions of photosynthesis. Stacks of thylakoids form the grana of chloroplasts.

## Tight Junction

Cell-cell junction that seals adjacent epithelial cells together, preventing the passage of most dissolved molecules from one side of the epithelial sheet to the other.

## Tight-receptor Structures

A receptor structure in which a bound ligand of a particular kind is confined on all sides by repulsive interactions (note that favorable binding energies are compatible with repulsive forces). A tight-receptor structure discriminates strongly against all molecules larger than the target.

## Tissue Engineering

Generating tissue *in vitro* for clinical applications, such as replacing wounded tissues or impaired organs. A *cell culture* facility is required for cell harvest and expansion. The term "tissue engineering" was coined at an National Science Foundation (NSF)—sponsored meeting in 1987. At a later NSF-sponsored workshop, tissue engineering was defined as "...the application of principles and methods of engineering and life sciences toward fundamental understanding ...and development of biological substitutes to restore, maintain and improve human tissue functions." This definition is intended to include procedures where the biological substitutes are cells or combinations of different cells that may be implanted on a scaffold such as natural collagen or as synthetic, biocompatible polymers to form a tissue.

## Titanium Nanopositioner Stages

If the positioning accuracy must be maintained over a prolonged time and temperatures cannot be maintained accurately, use of alternative materials may provide the necessary positioning stability. Titanium's coefficient of thermal expansion ($0.9 \times 10^{-5}$ mm/mm per °C) represents a significant improvement over aluminum alloys and offers the additional benefit of higher strength –improving nanopositioner performance.

## Top-down Approach

The synthesis of a nanostructured material or a nanoscopic particle, nanocrystal by 'sculpting' it from the macroscopic piece of unstructured material (bulk). Common techniques used for this purpose are lithography, etching, mechanical milling, engineering using atomic force microscope (which can be used to 'scratch' the surface leaving nanoscopic channels and structures behind)...

## Top-down Nanotechnology

Engineers taking existing devices, such as transistors, and making them smaller. Top-down or mechanical nanotechnology will have the greatest impact on our everyday lives in the near future.

## Toughness

A measure of the amount of energy absorbed by a material as it fractures, indicated by the total area under the material's tensile stress-strain curve.

## Transcription

Conversion of genetic information from DNA to messenger RNA (mRNA).

## Transduction

The conversion of the signal to be measured into another, more easily accessible form. Source of energy for transmission of the sensor signal.

## Transduction Mode

How the sensor acquires the desired information from the material. In general, this parameter is an indication of the ability of the sensor signal to provide information regarding a material property or state of interest.

## Transhuman

Someone actively preparing for becoming posthuman. Someone who is informed enough to see radical future possibilities and plans ahead for them, and who takes every current option for self-enhancement.

### Transhumanism

Philosophies of life (such as Extropianism) that seek the continuation and acceleration of the evolution of intelligent life beyond its currently human form and human limitations by means of science and technology, guided by life-promoting values.

### Transient Response

The response of the sensor to a step change in the measurand.

### Transistor

The transistor is a solid state semiconductor device that can be used for amplification, switching, voltage stabilization, signal modulation and many other functions. It allows a variable current, from an external source, to flow between two of its terminals depending on the smaller voltage or current applied to a third terminal. Transistors are made either as separate components or as part of an integrated circuit.

Transistors are divided into two main categories: *bipolar junction transistors* (BJTs) and *field effect transistors* (FETs). Transistors have three terminals where, in simplified terms, the application of current (BJT) or voltage (FET) to the input terminal increases the amount of charge in the active region, and thereby effectively increases the conductivity between the other two terminals, and hence controls current flow between those terminals. The physics of this "transistor action" is quite different for the BJT and FET.

In analog circuits, transistors are used in amplifiers, (direct current amplifiers, audio amplifiers, radio frequency amplifiers), and linear regulated power supplies. Transistors are also used in digital circuits where they function as electrical switches. Digital circuits include logic gates, random access memory (RAM), and microprocessors.

The first patents for the transistor principle were registered in Germany in 1928 by Julius Edgar Lilienfeld. In 1934 German physicist Dr. Oskar Heil patented the field-effect transistor. It is not clear whether either design was ever built, and this is generally considered unlikely.

On 22 December 1947 William Shockley, John Bardeen and Walter Brattain succeeded in building the first practical point-contact transistor at Bell Labs. This work followed from their war-time efforts to produce extremely pure germanium "crystal" mixer diodes, used in radar units as a frequency mixer element in microwave radar receivers. Early tube-based technology did not switch fast enough for this role, leading the Bell team to use solid state diodes instead. With this knowledge in hand they turned to the design of a triode, but found this was not at all easy. Bardeen eventually developed a new branch of surface physics to account for the "odd" behaviour they saw, and Bardeen and Brattain eventually succeeded in building a working device.

Bell Telephone Laboratories needed a generic name for the new invention: "Semiconductor Triode", "Solid Triode", "Surface States Triode", "Crystal Triode" and "Iotatron" were all considered, but "transistor," coined by John R. Pierce, won an internal ballot.

Bell put the transistor into production at Western Electric in Allentown, Pennsylvania. They also licensed it to a number of other electronics companies, including Texas Instruments, who produced a limited run of transistor radios as a sales tool. Another company liked the idea and also decided to take out a license, introducing their own radio under the brand name Sony. Early transistors were "unstable" and only suitable for low-power, low-frequency applications, but as transistor design developed, these problems were slowly overcome. Over the next two decades, transistors gradually replaced the earlier vacuum tubes in most applications and later made possible many new devices such as integrated circuits and personal computers. Shockley, Bardeen and Brattain were honored with the Nobel Prize in Physics "for their researches on semiconductors and their discovery of the transistor effect". Bardeen would go on to win a second Nobel in physics, one of only two people to receive more than one in the same discipline, for his work on the exploration of superconductivity.

In August 1948 German physicists Herbert F. Mataré (1912– ) and Heinrich Walker (ca. 1912–1981), working at Compagnie des Freins et Signaux Westinghouse in Paris, France applied for a patent on an amplifier based on the minority carrier injection process which they called the "transistron."

Since Bell Labs did not make a public announcement of the transistor until June 1948, the transistron was considered to be independently developed. Mataré had first observed transconductance effects during the manufacture of germanium duodiodes for German radar equipment during WWII.

## Transition State

At the saddle point of a col linking two potential wells, the direction of maximum negative curvature defines the reaction coordinate; the transition state is a hypothetical system of reduced dimensionality, free to move only on a hypersurface perpendicular to the reaction coordinate at its point of maximum energy.

## Translation

Conversion of genetic information from messenger RNA (mRNA) into protein.

## Transmission Electron Microscopy

Any technique in which an electron transparent sample is bombarded with an electron beam and the intensity of the transmitted electrons which is determined by scattering phenomena (electron absorption phenomena) in the interior of the sample is recorded. TEM essentially provides a high resolution image of the microstructure of a thin sample. This technique is often just called *electron microscopy*. The term transmission electron microscopy is however recommended for the sake of a clear distinction from other electron microscopic techniques.

# U

## Unit Cell

The crystal structure of a material or the arrangement of atoms in a crystal structure can be described in terms of its unit cell. The unit cell is a tiny box containing one or more motifs, a spatial arrangement of atoms. The units cells stacked in three-dimensional space describes the bulk arrangement of atoms of the crystal. The unit cell is given by its lattice parameters, the length of the cell edges and the angles between them, while the positions of the atoms inside the unit cell are described by the set of atomic positions $(x_i, y_i, z_i)$ measured from a lattice point.

Although there are an infinite number of ways to specify a unit cell, for each crystal structure there is a *conventional unit cell,* which is chosen to display the *full symmetry* of the crystal (see below). However, the conventional unit cell is not always the smallest possible choice. A primitive unit cell of a particular crystal structure is the smallest possible volume one can construct with the arrangement of atoms in the crystal such that, when stacked, completely fills the space. This primitive unit cell does not always display all the symmetries inherent in the crystal. A Wigner-Seitz cell is a particular kind of primitive cell which has the same symmetry as the lattice. In a unit cell each atom has an identical environment when stacked in 3 dimensional space. In a primitive cell, each atom may not have the same environment.

## Unsaturated

Describes a molecule that contains one or more double or triple carbon-carbon bonds, such as isoprene or benzene. Possessing double or triple bonds.

## Utility Fog

Utility fog is a collection of tiny robots, envisioned by Dr. John Storrs Hall while he was thinking about a nanotechnological replacement for car seatbelts. The robots would be microscopic, with extending arms reaching in several different directions, and can perform lattice reconfiguration. Grabbers at the ends of the arms would allow the robots (or foglets) to mechanically link to one another and share both information and energy, enabling them to act as a continuous substance with mechanical and optical properties that could be varied over a wide range.

Each foglet would In the original application as a replacement for seatbelts, the swarm of robots would be widely spread-out, and the arms loose, allowing air flow between them.

In the event of a collision the arms would lock into their current position, as if the air around the passengers had abruptly frozen solid. The result would be to spread any impact over the entire surface of the passenger's body. Utility fog is sometimes thought of as a nanotechnological version of the Swiss Army Knife.

While the foglets would be micro-scale, construction of the foglets would require full molecular nanotechnology. Each bot would be in the shape of a dodecahedron with 12 arms extending outwards. Each arm would have 4 degrees of freedom. When linked together the foglets would form an octet truss. The foglets' bodies would be made of aluminum oxide, not diamond, to avoid creating a fuel air explosive.

# V

## Valence Electrons

In chemistry, valence electrons are the electrons contained in the outermost, or *valence*, electron shell of an atom. Valence electrons are important in determining how an element reacts chemically with other elements: The fewer valence electrons an atom holds, the less stable it becomes and the more likely it is to react. The reverse is also true, the more full/complete the valence shell is with valence electrons, the more inert an atom is and the less likely it is to chemically react with other chemical elements or with chemical elements of its own type. This is because it takes more transfer of energy (photons) to lose or gain an electron from or into a shell when that shell is more complete/full.

Valence electrons have the ability like electrons in inner shells to absorb or release energy(photons). This gain or loss of energy can trigger an electron to move/jump to another shell or even break free from the atom and its valence shell. When an electron absorbs/gains more energy(photons), then it moves to a more outer shell depending on the amount of energy the electron contains and has gained due to the absorption of 1 or more photons. (Also see: electrons in an excited state)

When an electron releases/loses energy(photons), then it moves to a more inner shell depending on the amount of energy the electron contains and has lost due to the release of 1 or more photons.

## Vasculocyte

In medical nanorobotics, a theorized (nanorobotic) device

capable of performing repairs of an injured vascular luminal surface.

## Very large scale integration

Very-large-scale integration (VLSI) is the process of creating integrated circuits by combining thousands of transistor-based circuits into on a single chip. VLSI began in the 1970s when complex semiconductor and communication technologies were being developed.

The first semiconductor chips held one transistor each. Subsequent advances added more and more transistors, and as a consequence more individual functions or systems were integrated over time. The microprocessor is a VLSI device.

The first "generation" of computers relied on vacuum tubes. Then came discrete semiconductor devices, followed by integrated circuits. The first Small-Scale Integration (SSI) ICs had small numbers of devices on a single chip — diodes, transistors, resistors and capacitors (no inductors though), making it possible to fabricate one or more logic gates on a single device. The fourth generation consisted of Large-Scale Integration (LSI), i.e. systems with at least a thousand logic gates. The natural successor to LSI was VLSI (many tens of thousands of gates on a single chip). Current technology has moved far past this mark and today's microprocessors have many millions of gates and hundreds of millions of individual transistors.

As of mid-2004, billion-transistor processors are not yet economically feasible for most uses, but they are achievable in laboratory settings, and they are clearly on the horizon as semiconductor fabrication moves from the current generation of 90 nanometer (90 nm) processes to the next 65 nm and 45 nm generations.

At one time, there was an effort to name and calibrate various levels of large-scale integration above VLSI. Terms like Ultra-large-scale Integration (ULSI) were used. But the huge number of gates and transistors available on common devices has rendered such fine distinctions moot. Terms suggesting more-than-VLSI levels of integration are no longer in widespread use. Even VLSI is now somewhat quaint, given the common assumption that all microprocessors are VLSI or better.

### Virtual Microscope

The Virtual Microscope project is an initiative supervised by Rutgers University to make micromorphology and behavior of some small organisms explorable online. The server is located in the northern USA and the images are from Antarctica and the Baltic Sea. The user can click into the details of interest and get images of higher magnification or resolution, increasing to Scanning Electron Microscopy images, Transmission Electron Microscopy images, and at the end, peer reviewed publications.

On larger objects the arrow keys of the keyboard can be used as the crosstable on a real microscope is used. This project also introduces young learners to the internet and the capabilities of web browsers, search engines and databases. The editor board consists of professors from universities on both sides of the Atlantic.

### Viscosity

The ratio of the magnitude of an applied shear stress to the velocity gradient that it produces; in other words: a measure of a noncrystalline material's resistance to permanent deformation.

### Visible Spectrum

Visible light makes up only a small part of the *electromagnetic spectrum*. The visible light spectrum can be divided into different wavelenghts of light. The wavelength of the light determines the color of that light. The light spectrum goes from violet to red where red is the longest wavelength.

### Vitamins (Engineering)

In machine replication theory, vitamin parts are components of a self-replicating machine which the machine is incapable of producing itself, therefore these vital parts must be supplied from an external source.

# W

## Weather Satellite

A type of satellite used to give meteorologists information about the weather. Weather satellites can do things like take pictures of cloud cover, monitor threatening weather systems like hurricanes, measure temperatures of the air and the sea, and generally give meteorologists the information that they need to make weather predictions.

## Weber

Unit of magnetic flux. One weber is a magnetic flux that, linking a circuit of 1 turn, would produce in it an electromotive force of 1 volt if it were reduced to zero at a uniform rate in 1 second.

## Wet Nanotechnology

The study of biological systems that exist primarily in a water environment. The functional nanometre-scale structures of interest here are genetic material, membranes, enzymes and other cellular components.

The success of this nanotechnology is amply demonstrated by the existence of living organisms whose form, function, and evolution are governed by the interactions of nanometre-scale structures.

## Whisker

An old crystal radio receiver would often use a thin wire called a "cat's whisker" as a contact on a crystal of galena to form a diode.

## White Blood Cell (leucocyte)

White blood cells (a.k.a. leukocytes) are cells which form a component of the blood. They are produced in the bone marrow and help to defend the body against infectious disease and foreign materials as part of the immune system. There are normally between $4x10^9$ and $11x10^9$ white blood cells in a litre of healthy adult blood - about 7,000 to 25,000 white blood cells per drop. In conditions such as leukemia this may rise to as many as 50,000 white blood cells in a single drop of blood. As well as in the blood, white cells are also found in large numbers in the lymphatic system, the spleen, and in other body tissues.

# X

## Xenotransplantation

Xenotransplantation involves the use of live animal cells, tissues and organs in the treatment or mitigation of human disease. Xenotransplantation has the potential to provide an alternative source of tissue for human transplantation. However, there are concerns that in crossing the species barriers, infectious agents which normally do not cause any problem in the original species may be changed or may adversely respond to the new species into which the infectious agent is transplanted. Immune rejection remains the biggest challenge for xenotransplantation. The problem exists even for human to human transplants (known as allotransplantation), but is more serious for transplants between different species. Nearly all mammalian cells have markers which enable the immune system to recognise them not being foreign. The more different the genetic code between the donor organ and recipient, the greater the difference between a "self" marker and a "foreign" marker.

Some companies are currently developing transgenic animals such as pigs, that produce human markers. Cross-species transplants are more likely to produce host-vs-graft or graft-vs-host reactions than same-species transplants, because of the lack of antigenic similarity. Organisms which have been genetically engineered to reduce this lack of similarity have been produced but are not yet used to any significant degree in medical care. A worrisome element of xenotransplantation is the potential for infectious disease to spread from the donor animal, which is called xenozoonosis. One example is porcine endogenous retroviruses (PERVs) which are viruses within pigs that pigs are immune to, but can infect humans. Some recipients of pig neural cell transplants have had to agree to never donate blood, take frequent blood tests and use safe sex methods for the rest of their lives due to the risk of spreading such viruses. However, the patients who have received these pig cell transplants have yet to show any PERV-type infection.

## X-ray Crystallography

A technique in crystallography in which the pattern produced by the diffraction of X-rays through the closely spaced lattice of atoms in a crystal is recorded and then analyzed to reveal the nature of that lattice. This generally leads to an understanding of the material and molecular structure of a substance. The spacing in the crystal lattice can be determined using Bragg's law. The electrons that surround the atoms, rather than the atomic nuclei themselves, are the entities which physically interact with the incoming X-ray photons. This technique is widely used in chemistry and biochemistry to determine the structures of an immense variety of molecules, including inorganic compounds, DNA and proteins. X-ray diffraction is commonly carried out using single crystals of a material, but if these are not available, microcrystalline powdered samples may also be used, although this requires different equipment, gives less information, and is much less straightforward.

In inorganic chemistry, x-ray crystallography is used to determine lattice structures as well as chemical formulas, bond lengths and angles. The primary methods used in inorganic structures are powder diffraction and single-crystal diffraction. Many complicated inorganic and organometallic systems have been analyzed using single crystal methods, such as fullerenes, metalloporphyrins, and many other complicated compounds.

Single crystal is also used in pharmaceutical industry, due to recent problems with polymorphs. The major limitation to the quality of single-crystal data is crystal quality. Inorganic single-crystal x-ray crystallography is commonly known as small molecule crystallography, as opposed to macromolecular crystallography!

X-ray powder diffraction finds frequent use in materials science because sample preparation is relatively easy, and the test itself is often rapid and non-destructive.

The vast majority of engineering materials are crystalline, and even those which are not yield some useful information in diffraction experiments. The pattern of powder diffraction peaks can be used to quickly identify materials (thanks to the JCPDS pattern database), and changes in peak width or position can be used to determine crystal size, purity, and texture. The first protein crystal structure was of sperm whale myoglobin, as determined by Max Perutz and Sir John Cowdery Kendrew in 1958, which led to a Nobel Prize in Chemistry.

The X-ray diffraction analysis of myoglobin was originally motivated by the observation of myoglobin crystals in dried pools of blood on the decks of whaling ships. Today X-ray crystallography is used by pharmaceutical companies to determine specifically how drug lead compounds interact with their protein targets. Biological X-ray crystallography is to date the most prolific discipline within the area of Structural biology; out of the ~35000 protein structures solved, X-ray crystallography is responsible for ~29000. NMR spectroscopy has contributed almost 5000 and electron microscopy just over 100.

# Z

## Zener Diode

A Zener diode is a type of diode that permits current to flow in the forward direction like a normal diode, but also in the reverse direction if the voltage is larger than the rated breakdown voltage or "Zener voltage".

A conventional solid-state diode will not let significant current flow if reverse-biased below its reverse breakdown voltage. By exceeding the reverse bias breakdown voltage, a conventional diode is subject to high current flow due to avalanche breakdown.

Unless this current is limited by external circuitry, the diode will be permanently damaged. In case of large forward bias (current flow in the direction of the arrow), the diode exhibits a voltage drop due to internal resistance. The amount of the voltage drop depends on the design of the diode.

A Zener diode exhibits almost the same properties, except the device is especially designed so as to have a greatly reduced breakdown voltage, the so-called Zener voltage. A Zener diode contains a heavily doped p-n junction allowing electrons to tunnel from the valence band of the p-type material to the conduction band of the n-type material. A reverse-biased Zener diode will exhibit a controlled breakdown and let the current flow to keep the voltage across the Zener diode at the Zener voltage. For example, a diode with a Zener breakdown voltage of 3.2 V will exhibit a voltage drop of 3.2 V if reverse biased. However, the current is not unlimited, so the Zener diode is typically used to generate a reference voltage for an amplifier stage, or as a voltage stabilizer for low-current applications.

The breakdown voltage can be controlled quite accurately in the doping process. Tolerances to within 0.05% are available though the most widely used tolerances are 5% and 10%.

The effect was discovered by the American physicist Clarence Melvin Zener.

Another mechanism that produces a similar effect is the avalanche effect as in the avalanche diode. The two types of diode are in fact constructed the same way and both effects are present in diodes of this type. In silicon diodes up to about 5.6 volts, the zener effect is the predominant effect and shows a marked negative temperature coefficient. Above 5.6 volts, the avalanche effect becomes predominant and exhibits a positive temperature coefficient.

In a 5.6 V diode, the two effects occur together and their temperature coefficients neatly cancel each other out, thus the 5.6 V diode is the part of choice in temperature critical applications.

Modern manufacturing techniques have produced devices with voltages lower than 5.6 V with negligible temperature coefficients, but as higher voltage devices are encountered, the temperature coefficient rises dramatically. A 75 V diode has 10 times the coefficient of a 12 V diode.

All such diodes, regardless of breakdown voltage, are usually marketed under the umbrella term of 'zener diode'.

*Uses of Zener diodes:* Zener diodes are widely used to regulate the voltage across a circuit. When connected in parallel with a variable voltage source so that it is reverse biased, a zener diode conducts when the voltage reaches the diode's reverse breakdown voltage. From that point it keeps the voltage at that value.

## Zeptosecond

One-billion-trillionth of a second, or $10^{-21}$ second. Because nuclear movement takes place so quickly, scientists would need a pulse of light lasting just one zeptosecond to observe them. Johns Hopkins University

## Zero Sum

Zero-sum describes a situation in which a participant's gain or loss is exactly balanced by the losses or gains of the other participant(s). It is so named because when the total gains of the participants are added up, and the total losses are subtracted, they will sum to zero. Chess is an example of a zero-sum game - it is impossible for both players to win.

Zero-sum is a special case of a more general constant sum where the benefits and losses to all players sum to the same value. Cutting a cake is zero- or constant-sum because taking a larger piece reduces the amount of cake available for others.

Situations where participants can all gain or suffer together, such as a country with an excess of bananas trading with another country for their excess of apples, where both benefit from the transaction, are referred to as non-zero-sum. Other non-zero-sum games are games in which the sum of gains and losses by the players are always less than what they began with, such as in a game of poker played in a casino in which vigorish is taken by the house.

The concept was first developed in game theory and consequently zero-sum situations are often called zero-sum games though this does not imply that the concept, or game theory itself, applies only to what are commonly referred to as games. Optimal strategies for two-player zero-sum games can often be found using minimax strategies.

In 1944 John von Neumann and Oskar Morgenstern proved that any zero-sum game involving *n* players is in fact a generalised form of a zero-sum game for two persons, and that any non-zero-sum game for *n* players can be reduced to a zero-sum game for $n + 1$ players; the $(n + 1)$ player representing the global profit or loss.

## Zettatechnology

As the term nanotechnology has been adopted by the broad community to cover any development that utilises sufficient nanoscale features, Drexler argues that the term is misleading when applied to the